CREATING ISLAND RESORTS

Creating Island Resorts is a study of tropical island resorts, the areas they occupy, the people who live and work there and the tourists who visit them. An island resort is a special place – a pleasure periphery set apart from the mainland and from the mainstream – a community exhibiting many of the key characteristics of postmodernism. This book includes, but goes beyond the more commonly encountered marketing and economic analyses of resort destinations, by examining social, cultural, mythical, environmental, organisational and political dimensions.

The study offers a comparative analysis of two specific destination areas – the Mamanuca Islands in Fiji and the Whitsunday Islands in Queensland, Australia with special reference to the Australian market. The book highlights some of the special challenges facing island resort destinations in developing countries such as Fiji, relative to developed countries such as Australia and the differences and similarites between equivalent domestic and international resort destinations. The case material includes consumer focus groups in the key source markets, a detailed telephone survey of travel agents in the same places, and personal interviews with resort managers and with key stakeholders from the public and private sectors.

Brian E.M. King is an Associate Professor and Deputy Head of the Department of Hospitality and Tourism Management, Victoria University of Technology, Melbourne.

ROUTLEDGE ADVANCES IN TOURISM
Series editor: Brian Goodall

CREATING ISLAND RESORTS

Brian E.M. King

London and New York

First published 1997
by Routledge
2 Park Square, Milton Park, Abingdon, Oxon, OX14 4RN

Simultaneously published in the USA and Canada
by Routledge
711 Third Avenue, New York, NY 10017

Routledge is an imprint of the Taylor & Francis Group

First issued in paperback 2011

© 1997 Brian E.M. King

Typeset in Garamond by
J&L Composition Ltd, Filey, North Yorkshire

British Library Cataloguing in Publication Data
A catalogue record for this book is available from the British Library

Library of Congress Cataloging in Publication Data
A catalogue record for this book has been requested

ISBN13: 978-0-415-14989-1 (hbk)
ISBN13: 978-0-415-51357-9 (pbk)

Publisher's Note
The publisher has gone to great lengths to ensure
the quality of this reprint but points out that some imperfections in
the original may be apparent.

CONTENTS

CONTENTS

FIGURES

TABLES

ACKNOWLEDGEMENTS

Many people helped me with the preparation of this book and with the doctoral thesis upon which it is based.

I would like to thank my colleagues at Victoria University of Technology: Ari Gamage, Deirdre Giblin, Leo Jago, Liz Hillary, Ian Priestly, Robin Shaw, Ken Wilson, Brian Wise, Austin Norman, Thomas Bauer, Julie Stuart, Anne Marie Kemp, Gerald Sullivan, Sherill Pointon, Briony Barker and Vanessa Riedl for their help and encouragement. Paul Whitelaw deserves special mention for his assistance with data analysis.

A number of people made helpful comments on aspects of my research or supplied me with useful information: John Arnold, Rene Baretje, Neil Black, Mark Bonn, Chris Cooper, Peter Dieke, Michael Fagence, Justin Francis, Brian Goodall, Geoff Hyde, Barry Jones, Kiki Kaur, Neil Leiper, Mike McVey, Rik Medlik, Ady Milman, Douglas Pearce, Abe Pizam, Dorasamy Rao, John Rickard, Tony Seaton, Sean Weaver, Jeff Wilkes and Professor Paul Wilkinson.

Staff at Brian Sweeney and Associates provided me with valuable research assistance and generous access to their resources. Thanks to Brian Sweeney, Peter Hennessy, Lorraine Merry and Julie Pirotta. I am especially grateful to Mike Edwardson for his insights, support and friendship.

The tourism industry representatives whom I interviewed are listed in the appendix which precedes the bibliography. A number of others provided me with assistance outside the formal interview process and offered me an industry perspective. These included Alistair MacIntyre, Bob Walker, Col Willett, Isimeli Bainimara, Margaret Thaggard, Rob Lindsay, Martin McKinnon, Stephen Gregg and Tracey Walker. Special thanks to Dick and Carole Smith for their fine hospitality at Musket Cove.

Hillary and Wendy Ducros, Sam Oliphant and Matthew Smith all helped me with editing and document preparation.

Special thanks to Peter Spearritt for his patience, insights, enthusiasm and guidance.

Finally, I would not have persevered without the constant support and understanding of my loving partner, Judy. Our daughter Mimi made sure that I did not lose sight of the important things in life.

1

INTRODUCTION

THE MYTH OF THE TROPICAL ISLAND AS AN 'EARTHLY PARADISE'

To the European mind, tropical islands are bountiful places offering travellers an escape from everyday realities and a temporary materialisation of what is imagined to be the 'good life'. The jet aircraft has brought island paradises tantalisingly close for the relatively affluent of the world. For North Americans the Caribbean is delightfully close as are the islands of the Indian Ocean for more affluent South Africans. Despite their undoubted appeal, neither the Caribbean nor the Indian Ocean have quite the intoxicating attraction of the South Pacific. These previously most inaccessible of tropical islands now face the challenges of adapting to an influx of pleasure-oriented travellers from throughout the world, or in the case of those which have not yet established themselves as tourism destinations, to solicit such an influx.

The Pacific islanders who inhabit the 'paradise' to which so many Europeans have aspired are also under pressure to adapt to the temporary holiday migrations from across the globe. Australians have long regarded the South Pacific as their 'back-yard' and the South Pacific 'paradise' has always been more accessible for them than for residents of the Northern Hemisphere. Australian travellers are having to share Pacific holiday resorts with visitors from throughout the world. But do Australians really need to travel overseas to find the type of paradise so deeply embedded in the western imagination? Australia's own Great Barrier Reef is fringed by numerous tropical islands. Does this imply that paradise may be experienced without the need for a passport? Do Australians weigh up the prospect of travel to Queensland with its connotation of patriotism and of 'buying Australian' against the option of an overseas trip? Might both destinations lose their allure because paradise has become too easily accessible, too prone to excessively hyperbolic promotion and too easy to compare image with reality?

Tourism paradises have been described as 'the most exploited commonplace in international advertising' (Giesz 1968: 103). European ideas of an

1

earthly paradise predated the discovery of the South Pacific islands by explorers from the Northern Hemisphere. According to Bloch (1959 vol. 2: 890) 'Dante, assuming a spherical form of the earth, even relegated it [paradise] to the *antipodes* of Jerusalem, into the South Seas'. Another example is the Polynesian Hawaiki referred to by Cohen (1982). Cohen maintains that the idea of paradise was merged into the myth of a South Continent or *Terra Australis Incognita*. According to Eliade 'the myth of the Earthly Paradise has survived until today in adopted form as an "Oceanic paradise"' (Eliade 1952: 11–12). Niederland (1957: 56) observes that 'the site of the original Earthly Paradise has often been imagined to be an island' and Turner and Ash suggest that 'the most potent images of man's former idyllic natural state are located on islands: the homelands of the noble savage, the original sites of the Garden of Eden' (1975: 151).

The tropical island embodies the central holiday aspirations of western consumers. The image of tropical islands as carefree, romantic and adventurous places is a result of literary associations supplemented by deliberate tourism marketing through the twentieth century, some of it undertaken by Pacific islands. Children's novels such as Ballantyne's *Coral Island* (1858), Stevenson's *Treasure Island* (1883) and Jack London's *South Sea Tales* (1911) have influenced impressionable readers by creating an association between adventure, self-discovery and the tropical South Pacific island. Such impressions have been perpetuated into adulthood by literature such as *By Reef and Palm* (1894), in which Louis Becke flits between various Pacific islands and in the post-war era by Michener's *Tales of the South Pacific* (1947) and *Return to Paradise* (1951). Cohen (1982) has argued that the writings of Mark Twain and Jack London were partly responsible for the opening up of Hawaii to tourism. The visual arts such as Gauguin's paintings of Tahiti have also played a role in the spread of Pacific imagery, by portraying indigenous people in exotic landscape settings. Michener's work is perhaps best known in its Rogers and Hammerstein musical version *South Pacific* which is frequently performed in Australia. The popular British BBC radio programme *Desert Island Discs* has added an extra musical dimension to the tropical island myth. In that show, well-known interviewees select the eight pieces of music that they would most like to accompany them if they were stranded on a desert island. Advertising has also played its part. A great deal of confectionery and soft drinks advertising is staged in tropical settings. There are examples of references and cross-references to the Pacific tropical island myth throughout the developed world.

Desert island images are not always positive. Defoe's *Robinson Crusoe* (1719–20) evokes in its readers, a fear of abandonment in an uninhabited place. In Golding's *Lord of the Flies* (1954) the island setting is devoid of the attributes of Western 'civilisation', and the group of boys marooned there soon fall victim to rivalry and violence in the absence of conventional social controls. Despite these menacing connotations which often lurk in the subconscious, the positive images have been systematically and selectively built upon by tourism marketeers: promotions of Tahiti frequently refer to the legacy of Gauguin, and promotions for Western Samoa lay claim to Stevenson. The blend of literary and artistic images with recent and more blatantly commercial imagery, has reinforced the view of tropical islands as paradise and has played down negative connotations such as isolation and the absence of metropolitan comforts and of social controls. In the tourism marketing literature, different labels (Gunn 1988a) are ascribed to the two essential types of imagery. Imagery deliberately intended to increase visitation to a destination is described as 'induced', whilst the complex array of images which affect people's impressions of a destination, but are not purposefully commissioned by tourism agencies, are called 'organic' images. Information gathering plays a key part in the process of choosing a destination with a significant role being played by word-of-mouth recommendation from friends or relatives. Such recommendations are particularly important for first-time visitors.

A curious element of island resort tourism is that without a powerful dose of what Hall has called 'boosterism', it might scarcely exist (1991). The language of exaggeration and excess deployed in advertising, brochures and other promotional material is used frequently to boost all forms of tourism demand. In the case of tropical island resorts, such language is critical to the creation of demand at its most fundamental level. Why are island resort consumers so prone to being seduced by the promised gratifications of boosterism? Firstly, few are aware of the true conditions prevailing in island resorts prior to their departure from home. Repeat visitors – who have some destination knowledge – make up only a small minority of total island resort visitation and consumers rarely have the opportunity to sample the place before making a major financial outlay. It is common practice in other industries – but not in tourism – to allow consumer sampling of a product prior to purchase. The practice is even less common in the case of island resorts. Since most tropical islands are remote, travellers rarely encounter them *en route* to other main destinations. Applying the typologies developed by Leiper as part of his 'tourism system', it is clear that most tropical islands function as destination regions and not as transit regions (Leiper 1990a) and depend on consumers having a strong disposition to travel to that particular destination. Boosterism aside, there is a significant group of consumers who return repeatedly to a particular resort, apparently untroubled by any inconsistency between image and reality.

Consumer perceptions owe more to the mythology of the palm-fringed coral atoll, than to the realities of life in the tropics. In tourism promotions, water is seen as embodying purity, romance, nature and the pleasurable elements, whilst at the same time symbolising luxury through its manipulation by contemporary technology into giant swimming pools, spas, waterfalls and other configurations. Less conveniently for the positive imagery, tropical water is periodically associated with sharks, stonefish, sea wasps, coral cuts and may prompt sunburn. The summer months in tropical resorts usually coincide with the cyclone season when much of the annual rainfall occurs. Despite this, holidaymakers appear willing to surrender themselves to the powerful allure of the tropical island idea. They willingly opt for isolated resorts where, in reality, they are usually compelled to purchase goods and services from a single operator whilst they expect a full range of western style conveniences and facilities to be provided. If poor weather or some other influence leads to disappointment and they choose to depart early, they sacrifice a large part of their investment. In practice few leave early, choosing instead to submit themselves to a suboptimal experience. In some respects, island resort holidaymakers are amongst the most gullible travellers, willingly submitting themselves to the tyranny of the resort corporation, in search of an experience that, they might grudgingly acknowledge, could be an illusion.

Taking the imagery of tropical islands as a point of departure, it is clear that island resorts can be conceptualised as undergoing a process of creation (in the popular imagination) and subsequently of re-creation as more blatantly promotional images come into play. This study compares how two island resort groups have been created and re-created. In addition to the perceptual dimension where images are formulated and subsequently projected onto potential consumers, there is also a physical setting. Resorts and the infrastructure which accompanies them are firstly conceived and built and in many cases, later re-built. In the present study two destination regions are examined – the Whitsunday Islands in North Queensland, Australia and the Mamanucas in Fiji. The destinations are evaluated in relation to a particular source market, namely Australia's most populous states of New South Wales (NSW) and Victoria and their state capitals of Sydney and Melbourne in particular.

This study draws upon a variety of disciplines including sociology, history, anthropology, politics, environmental science and marketing. The social, cultural, environmental and political dimensions of island tourism are examined both from the destination perspective and from the perspective of the tourism-generating region. The study examines a particular source market and the methods by which the two regions can make themselves attractive to prospective consumers and travel agents. Poon's concept of the 'new tourism' is examined for its relevance to the holiday visitors who frequent the two destinations (Poon 1993). The study evaluates the social, cultural and environmental context of the resorts, acknowledging the commercial imperative to remain attractive to key source markets. The study is not prescriptive and recognises that different styles of destination will appeal to different consumers. Nevertheless, it is hoped that common issues and experiences will emerge from two regions manifesting these different styles, through the pursuit of issues with potential validity for resort destinations generally.

The selection of one domestic destination (the Whitsundays) and one international destination (the Mamanucas) was deliberate. From the point of view of the generating area, both can be termed 'short-haul sun, sea and sand' destinations. By applying this broad categorisation to the two areas, we can compare and contrast them with equivalent destinations throughout the world. As previously stated, each major source market in the developed world may be regarded as having its own 'pleasure periphery' of short-haul holiday destinations (Turner and Ash 1975). The selection of one international and one domestic destination offered the prospect of challenging a number of assumptions about the perceived differences between the practice of domestic and international travel. In explaining how resorts are created, we look at how the various attributes of an island resort holiday are packaged into an individual resort concept and then into a regional concept. Understanding various attitudes and perceptions towards resorts was considered particularly important because resorts are fundamentally human 'creations', albeit dependent on their natural settings. The question of what constituted the 'resort experience' was addressed through an exploration of attitudes and perceptions. These were gathered through surveys undertaken with consumers, travel agents, tour operators, resort managers and destination marketing organisations.

To what extent are decisions about travel to tropical island resorts based on price and the availability of certain facilities? Or are they emotion-laden and driven by imagery, indicating that travel to resorts is an example of hedonic consumption? To what extent are images of the two destinations contrasting, with Australia regarded as a familiar developed country and the South Pacific regarded as 'undeveloped' but exotic? To what extent are the two destinations regarded as equivalent, with both offering highly substitutable island holidays in the sun?

In addition to examining perceptions of the destinations in key source markets, the book sets out to profile the type of tourism activity occurring in the islands and the way it is managed. The interviews with resort managers provide insights into the operations of individual properties. They also provide an indication of the level of regional co-operation which has occurred. The relationship between the islands and the mainland which adjoins them is a critical issue here. To what extent is the major decision-making concerning management and marketing undertaken on the islands and to what extent on the mainland? The relationship between the Mamanucas and Fiji as a whole, and between the Whitsundays and Queensland as a whole, is also relevant.

Both island groups have been significantly affected by issues of airline regulation. These issues are examined in the context of how the tourism industry is structured. To what extent have the different structures which have applied to domestic and outbound travel in Australia impacted upon the two destinations? Two other regulatory issues worthy of comparative analysis are labour relations and environmental control. To what extent have regulations affected employment structures in the two groups, or is remoteness from the major population centres a more critical issue? Has the imposition of environmental regulation differed between the two regions and has it had an impact upon the management of the resorts?

A number of social and environmental issues are examined. Is the presence (in the case of Fiji) or absence (in the case of the Whitsundays) of an indigenous population in the resort regions, an issue for tourism marketing? Will it become more so in future? To what extent can we regard island resorts as communities in their own right and to what extent are they more likened to large hotels supplemented with certain ancillary facilities and activities? To what extent does the marketing of the two regions give a true perspective of the environmental features of the destinations? Are such issues of concern to consumers currently or are they likely to become so in future? Are the two regions adapting quickly enough to changing market conditions? Are they enhancing their distinctiveness and/or competitiveness through an acknowledgement of regional cultural and environmental dimensions?

Who was responsible for developing tourism and for creating the various resorts? Was it brought about by government activity or by entrepreneurial endeavour? What has been the role of the local communities and landowners in the tourism development process? Is there a distinct identity and cultural heritage in the two regions? If so, is this reflected in the type of tourism activity offered? In an era where the word *ecotourism* is sometimes regarded as much that is best in tourism, is the natural environment in good hands? Does the fact that Fiji is a developing country and Australia a developed country have any impact on the approach taken? Are Australians interested in the environment as a part of the holiday experience? To what extent is the environment purely a 'backdrop' to island and resort locations? This book attempts to answer these questions.

The decision to incorporate consideration of both cultural and natural heritage issues is deliberate. In a South Pacific context, the indigenous population are inextricably linked with concepts of land and environment. Separating cultural and natural heritage is impratical. Most definitions of ecotourism do in fact incorporate a cultural heritage dimension (Ceballos-Lascurain 1991; Ecotourism Association of Australia 1992; Figgis 1992; Young 1992), though the concept of 'endemic tourism' (Oelrichs 1992) is probably more emphatic still in outlining the inextricable link between natural and cultural heritage.

INTRODUCTION

WHY A COMPARATIVE STUDY?

The value of comparative tourism studies has been outlined by Douglas Pearce, a New Zealand-based geographer and one of the most frequently cited authors in the tourism field (1993). According to Pearce, comparative studies 'offer tourism researchers a way forward in a field still largely dominated by descriptive, ideographic work' (1993: 32). Despite the potential value of comparative work, it has been observed that such studies are rare: 'tourism literature contains few international comparative tourism studies' (Dieke 1993). Many studies have focused on a particular developing country or on developing countries in general, but most have lacked a comparative framework (Jenkins 1980). Literature comparing developed and developing countries is even scarcer than studies comparing developing countries (Pizam, Milman and King 1994). The latter study was unusual in that it examined specific regions within a developing country (Nadi in Fiji) and a developed country (Central Florida, USA). In his comparative study of tourism in the Gambia and Kenya, Dieke notes that the absence of comparative studies is attributable to logistical, financial and methodological problems: 'Such studies are difficult to conduct, not only in terms of resources, but in the selection of variables and issues which will be accurately compared' (1993: 2).

Despite his enthusiasm for comparative work, Pearce points out some of the same methodological traps mentioned by Dieke (Pearce 1993). He stresses that effective comparative work should involve 'more than the mere juxtaposition of case studies, for to be comparative the analysis must at the very least draw out and attempt to account for similarities and differences' (1993: 21). Pearce assesses comparative studies on the basis of whether their results can be generalised and whether they help build and test theories.

A recent study, *Keeping Australians at Home: Tourism Import Replacement Analysis* published by the Bureau of Tourism Research in Canberra (Haigh 1993), compared the factors determining domestic versus international travel by Australians. The study focused on demand factors and did not compare particular destinations. It identified opportunities and impediments under the following headings: economic factors, travel factors and psychological factors. A number of these are examined in more detail in the present study, using empirical methods. An example is Haigh's assertion that the travel agent commission rates which apply in Australia are smaller for domestic travel (5 per cent) than for international travel (9 per cent) (1993: 42).

Until recently, the most developed forms of tourism have been associated with developed countries, with infrastructural provision justified by the needs of the domestic tourism market. Though many developing countries are now highly dependent on tourism, most of their tourism-specific infrastructure has been built recently, in many cases since 1950. Can the developing countries learn from the experience of the developed countries? In the nineteenth century, industrialising countries such as Germany and the United States learnt some lessons from the earlier industrial revolution experience of Britain and from the associated development and marketing of spas and seaside resorts. In a similar fashion, countries and regions coming comparatively later to tourism development can learn from the experiences of pioneer destinations. In some cases, the more developed destinations may also learn from the emerging destinations. This is particularly the case where regions are competing for similar markets and need to identify an appropriate market positioning, based on their relative strengths. Are destinations in developing countries better able to adapt to changing consumer demands (for example, higher expectations of 'environmental friendliness') because of the generally smaller scale of the resorts found there? Conversely, are developed destinations better able to adapt because they have easier access to financial and human resources? These are some of the questions which a comparative study can help answer.

With a view to giving greater precision to the conclusions of this study, the number of variables has been kept to a minimum. This approach also improves the prospect of realising the objectives that Pearce regarded as well served by the comparative approach. Limiting the number of destinations to two and the source market to one (i.e. Australia) allowed the author to achieve greater confidence in the research conclusions and go beyond the exploratory methods used in Haigh's study. The selection of two spatially compact destinations allowed a more detailed examination of destination development and marketing. At the same time both destinations are competitors for the same potential holidaymakers resident in south-eastern Australia (Melbourne and Sydney), thereby giving the study a dynamic element. The author did give some consideration to adding a third, South-East Asian resort destination to the study. This option was eliminated because of a concern that the direct comparison between domestic and international travel would become blurred. The potential cost of travel to multiple destinations by the author was also a constraint.

The Australian market was selected as a focus of the study for a number of reasons. The author is Melbourne-based and in view of the difficulties in researching a number of distant destinations (referred to by Dieke), it was more convenient and practical to focus on a readily accessible source market. Second, comparable sun, sea and sand destinations active within the selected source market were available. Third, the diversity of latitudes covered by the Australian continent (Australia is the only developed country occupying a whole continent) results in both cooler climate tourism generating regions (for example, the states of Tasmania and Victoria) and tropical/subtropical beach resort destinations (for example, the State of Queensland and the northerly part of Western Australia). As a result, much 'sunlust' tourism is domestic, unlike in Europe where mass sunlust tourism is predominantly outbound (Grey 1970) and international in character. This offers the possibility of arriving at an Australian perspective into the nature of short-haul sunlust tourism activity and an insight into how travel patterns may have an impact upon Australia's self-image within the Asia–Pacific region. The comparison of travel 'at home' and travel 'away from home' offers the prospect of insights into the nature of domestic and international destinations as perceived by tourists.

The vast scale of Australia has resulted in an awkward relationship with the tiny nations of the South Pacific (the novelist Paul Théroux described Australia as 'Meganesia' (1992)). The complex nature of the relationship has periodically been described as uninterested, paternalist and high-handed (Utrecht 1984; Crocombe 1987). Australia is sometimes accused of behaving in a neocolonial manner, often in its dealings with former colony Papua New Guinea, though it more typically portrays itself as an equal amongst truly independent countries. A comparative study focusing on the Australian market and on destinations both within Australia and within the South Pacific (Fiji) can provide insights into the relationships between generating tourism markets and comparable tourism destinations.

THE CHOICE OF STUDY REGIONS

In selecting tropical island destinations frequented by Australians, it was advisable to choose areas offering comparable accessibility, including length of flying time, recognition as established destinations within the Australian market and comparable consumer experiences. Queensland and Fiji both satisfied these requirements and could reasonably be classified as sun, sea and sand destinations. The Whitsunday and Mamanuca groups each offer equivalent island resorts, nine in the case of the Whitsundays, twelve in the case of the Mamanucas. Fiji was chosen because the author had already undertaken major tourism investigations there. The recent shift of emphasis by the Fiji Ministry of Tourism (FMOT) and the Fiji Visitors Bureau (FVB) away from marketing Fiji simply as a resort experience in Australia, to its current emphasis on the diversity of the tourism product focusing on environmental and cultural experiences was also a motivation, since it highlighted the sometimes conflicting demands of resort-based tourism and the need to develop the social and environmental context. The change of focus by government was in part a response to reports highlighting perceived traveller dissatisfaction with the resorts in Fiji (Stollznow Research 1990, 1992). The author was concerned to identify the likely repercussions of the change of emphasis on Fiji's longest-established island resort region, the Mamanucas. Many destinations are currently evaluating the extent to which traditional sun, sea and sand tourism should be superseded by a more environmentally-based style of tourism – often called ecotourism. The present study offered the prospect of showing how two equivalent destinations have dealt with such issues, both consciously and unconsciously.

The nature of previous research work on tourism in the Queensland resort islands also provided an impulse for the current research. Substantial work had been undertaken for the Queensland Tourist and Travel Corporation (QTTC) on consumer attitudes to the island resorts, but much of the research material has been intended for internal QTTC use and the rest has been given only limited exposure in the public domain. A scholarly reinterpretation of this research was deemed to be of potential value. Providing a comparative and detailed international comparison at regional level (the Mamanucas and the Whitsundays) could help transform highly localised research, enabling the development of theory-based generalisations favoured by Pearce. The experience of the two island groups offers the prospect of being an instructive example for island resort operators in other parts of the world.

STUDYING RESORTS

The literature review incorporated in this introductory chapter is deliberately modest with a greater emphasis on literature reviews at the beginning of each subsequent chapter relevant to the particular focus of that section of the book. In this way acknowledgement could be given to the enormous range of literature behind key words such as tourism in developing countries, imagery, ecotourism, planning and development, regional marketing, tourism resorts, travel industry structure and tourism flows.

The book starts from the premise that research on resorts has been fragmented and very discipline specific. An interdisciplinary approach is used to examine the role and function of island resorts from a sociocultural, political and environmental perspective and to relate this broad perspective to a particular market. The primary research component enriches the perspective by depicting the way in which various industry and consumer representatives conceptualise the resorts. The broad range of reference documents cited in the research helps provide a bridge between the often disparate perspectives on resorts by researchers who, according to Craik (1991), often fail to adopt a genuinely integrated approach.

Much has been written on the nature, characteristics and scale of tourism in developing countries (Jenkins 1982; Harrison 1992). A lot of this work evaluates relations between the principal tourism generating countries (predominantly the developed nations including Australia) and the world's tropical and subtropical destinations (predominantly in the developing countries, including Fiji). Such studies have included examinations of so-called 'centre–periphery relations' (Hoivik and Heiberg 1980), of the role of transnational corporations (Dunning and McQueen 1982a), studies of imagery (R.A. Britton 1979), the marketing of authenticity (Silver 1993) and of the role of the tour operator. Within this literature, there are a number of works evaluating tourism in small island destinations (Cazes 1987; Wilkinson 1994). Fiji, in particular, has been the subject of a number of such studies, notably by Britton, whose work emphasised the links between tourism, dependency and underdevelopment (Britton 1980; Britton 1982; Britton and Clarke 1987).

As a developed country, Australia has been less frequently evaluated from the point of view of tourism's negative impacts, though Craik (1991) has examined the negative dimensions of tourism development in Queensland, particularly in the islands. Barr (1990) examined the historical development of tourism in the Whitsunday islands, emphasising its increasing dependence on externally generated capital. Conscious of the paucity of existing studies comparing tourism regions in developing and developed countries, the author has set out to establish the scholarly and heuristic utility of a comparative approach.

There is a considerable body of work dedicated to the study of resorts. Since resorts often aspire to being self-contained tourism destinations in their own right, the literature has studied the phenomenon from a diversity of angles, not confined to the study of resort facilities *per se*. These approaches include resort development (Dean and Judd 1985; Stiles and See-Tho 1991), planning (R.A. Smith 1992), the assessment of local attitudes (Witter 1985), marketing (King and Whitelaw 1992), the resort life cycle (Butler 1980), resorts as communities (Stettner 1993), architecture (England 1980), landscaping (Ayala 1991a, 1991b) and key success factors (Wober and Zins 1995). A significant dimension of this literature is the role of resorts as 'enclave' developments, separated from the reality of daily life in adjacent areas or regions (Freitag 1994).

With the exception of Barr's historical examination of the Whitsundays, neither the Whitsundays nor the Mamanucas have been the subject of any dedicated studies. More general studies on Queensland (Brian Sweeney and Associates 1991) and on Fiji (Stollznow Research 1990, 1992) have provided insights into the perceptions of the two areas by consumers. In both cases the research was commissioned by the relevant tourism authorities (the QTTC and FVB respectively) and undertaken by a market research company. The fact that such studies are typically (and necessarily) prepared very quickly with scant attention to the relevant academic literature limits their use as sources. Nevertheless, they do provide a useful starting point for the current research, particularly in the absence of academic studies specific to the field of the comparative resort evaluation.

As a category of research, the study of resorts is a loosely defined field because the term itself sometimes refers to the very specific and sometimes to the very broad. The use of the term 'mega-resort' is an example of the first type. This expression is sometimes used to describe developments which satisfy a very specific criteria of high capital intensity. The term 'integrated resort' typically refers to properties which incorporate a wide range of recreational facilities and accommodation types (Styles and See-Tho 1991). When used in its broadest sense, the term 'resort' is used to describe a whole destination region – cities are sometimes described as resorts for example, notably in their promotional material, though in many cases the bulk of their visitation is accounted for by business travellers (who are classed as tourists in official definitions). According to this usage, virtually any destination can be called a resort.

For the purposes of the present research, the use of the term 'resort' to describe a complete destination region is not appropriate in the sense that one might describe Surfers' Paradise in Queensland as a resort. Whilst the various island resorts share collective interests and concerns, each is an 'integrated resort' in its own right. Their separation from the mainland by a stretch of water and their need to provide a full range of activities and facilities on site, qualifies them under this category, though in most cases their relatively small scale inhibits them from being mega-resorts. A key element of the present research examined the extent to which consumers hold a clearly defined picture of what constitutes a resort. Is the conceptual and technical debate over resort definitions symptomatic of confusion amongst consumers? To what extent are tropical islands seen as archetypes of the resort *genre*?

Any review of the research undertaken on resorts must acknowledge both the micro (specific) and the macro (general and broad-based) approaches noted earlier. It must also acknowledge the contribution to our understanding of the resort concept by scholars from diverse disciplines, attracted to the subject because it lends itself to both highly technical evaluation and to grand generalisation about holiday behaviour. Highly technical evaluations have been undertaken by specialists in business operations, construction, design and environmental planning to name but a few. The large-scale transnational investments attracted by many resorts have provided a justification for the attention of lawyers, financial planners, architects, environmental planners and designers who see potential profits and professional challenge in the field. A number of professionals in those fields have contributed to the literature.

Social scientists have also shown an interest in resorts, particularly but not exclusively at the macro level. The holiday resort epitomises the archetypal package tour-based tourism for many social scientists and provides an insight into leisure patterns. Resort-based tourism is seen as typifying many of the characteristics of modern life. Krippendorf describes resorts as 'therapy zones for the masses' referring to 'sun–sea therapy' at the coastal resorts and 'snow–ski therapy' in the mountains (1987). He refers to 'self-sufficient holiday complexes, designed and run on the basis of careful motivation studies as enclaves for holiday-makers. Total experience and relaxation. Fenced off and sterilised' (1987: 70–1). Resorts are a particular target for his acerbic analysis of contemporary tourism. He proposes a 'humanisation of travel' as a counterweight to some of the excesses experienced in resorts. Similar accusations of sterility are frequently levelled at contemporary shopping malls. Resorts share a common characteristic with shopping malls in that they are designated areas set aside for human consumption (Sack 1992). The concept and perception of resorts as consumption places is an issue that will be developed further in the present research.

The view that resorts embody many of the excesses of modern tourism has attracted the interest of geographers, social anthropologists, planners, sociologists, economists, historians and many others. The mere fact that the word 'resort' can be either a noun or a verb has been highlighted by some authors. Craik (1991) uses the word 'resort' in a pejorative way amounting to a term of abuse. Tourism is 'resorted to', in her opinion, only because no other viable alternative is available. In their book *The Earth as a Holiday Resort*, Boers and Bosch have lamented the increasing 'resortisation' of the world (1994). Which resort definitions can we apply equally in the Australia and the Fiji contexts? According to the Oxford Dictionary, a resort is 'a place resorted to, a popular holiday place'. Academic definitions of resorts have also tended to be general and somewhat descriptive and pragmatic. According to Gunn, resorts are 'complexes providing a variety of recreations and social settings at one location' (Gunn 1988a). An alternative definition (Burkart and Medlik 1981) is more specific in that it does refer to tourism, but it is still general: 'The term resort has come to acquire its literal meaning to denote any visitor centre to which people resort in large numbers' (1981: 45). The authors even encompass capital cities within this definition because such centres function 'as centres of commerce and government'!

Definitions of resorts which emphasise the commercial accommodation component are typically narrower, though not necessarily less confusing. In Australia, South stated that 'the method of classification of accommodation in Australia is continually debated and under review by the industry' (1988: 153). The motoring organisations are responsible for classifying accommodation. The Royal Automobile Club of Victoria (RACV) has defined a resort as somewhere 'offering extensive recreational facilities on the premises and [it] may cater to specific interests such as golf, tennis, fishing etc., with an all-inclusive tariff option' (RACV 1992: 6).

Resorts may be defined very narrowly (hotels offering a comprehensive range of recreational facilities on site) or very broadly (any recreational destination where travellers congregate, including capital cities). A useful summation of such diverse definitions is accommodated by Mieczkowski (1990). He bases his classification on an examination of various types of human settlement by Grunthal (1936). His three classifications included

13

settlements without tourism, settlements with tourism and finally what he describes as 'tourism settlements'. Mieczkowski's interpretation distinguishes between cities and resorts as the two key types of tourism settlements by applying two criteria, namely resources and function. He examines resorts in terms of a continuum extending from small scale to large scale. At one end he cites the example of a 100–200 room hotel and at the opposite end there are '"resort towns" or even "resort cities" of truly urban dimensions (Waikiki Beach, Hawaii)' (p.318). His definition goes some way to accommodating both the micro (by presenting a minimum size) and the macro. He does not, however, provide any categorical insights into the dilemma that the term 'resort' is sometimes used very loosely when referring to a tourism region. Using Mieczkowski's definition, all twenty-one resorts evaluated in the present research are clearly categorised as 'tourist settlements' since tourism is their sole *raison d'être*.

The debate about resort definitions in Australia progressed with the publication of the Australian Bureau of Statistics' *Standard Classification of Visitor Accommodation* (SCOVA) (Fleetwood 1993). For the present research, the author has modified SCOVA to arrive at a definition as follows:

A complex of tourism facilities incorporating accommodation, food and beverage and recreational provision located on an island. The destination is incorporated in air-inclusive tour programmes and offers the option of an inclusive package incorporating at least one meal per day plus other ancillary services. The destination offers a sufficient range of activities and attractions to lend itself to an extended length of stay, usually of between five and fourteen days.

The use of the term 'resort' by Australian properties is confused. A study by King and Whitelaw (1992), using a combination of the AAA classifications and how the properties chose to describe themselves, depicted the typical Australian resort as a three- and-a-half-star roadside motel in Northern NSW (King and Whitelaw 1992). In such (typical) cases, the 'resort' label is a 'booster' or lure to motorists and may be causing inconsistencies between expectations and actual holiday experiences. The problem may be less pronounced in the Whitsunday Islands where relative isolation necessitates the provision of resort facilities *in situ*, but it may be symptomatic of some confusion amongst Australian consumers as to what constitutes a resort and what to expect on arrival at the destination.

For the purposes of this study, the definition of resort is more prescriptive than many used elsewhere. On all of the islands examined in this study, tourism constitutes the only or the predominant activity. The resort guests do not need to share the island with other users with the exception of resort employees and their dependants and sometimes day trippers. Because tourism is the sole economic activity, any activities, facilities

14

and/or provisions needed or demanded by consumers, must be provided by the resorts themselves or by businesses involved in a direct commercial relationship with the resorts. To this extent, these resorts are 'integrated', though on a much smaller scale than the mega-resorts currently expanding in South-East Asia, often on the mainland. When viewed as providers of accommodation such island resorts are, in a sense, the antithesis of city-based properties. Their predominant market is leisure-based and not business-based. Their ethos is one of total relaxation, in contrast to central business district (CBD) properties, where the provision of leisure facilities is seen as an 'escape' from the business activities that involve guests during the day.

Probably the most significant promotional medium linking the resorts and potential consumers are travel brochures produced and distributed by tour operators and airlines. Because of the relatively standardised format of such brochures into a predetermined quantity of photographs and copy, comparison of different destinations is made simpler. This type of promotion is exemplified by the practice of Australia's largest tour operator and travel agency group, Jetset, which includes all of its South Pacific destinations in a single brochure labelled simply 'Resorts'. Does this place such destinations at a disadvantage against Asian countries where cultural and other distinguishing features are given greater prominence? Is this an impediment to groups of tropical islands trying to diversify their appeal? Are promotional media such as tour operators placing too much emphasis on the implicit clichés of the tropical island paradise and assuming that resort features are the main distinguishing features between different countries? Such resorts certainly devote most of their attention to highlighting resort facilities, rather than their destination attributes. Does the marketing of the Whitsundays also suggest that they are islands dedicated to resort life? Do descriptions of the destinations amount to little more than the component physical features of the resort buildings? Is the classification and description of resorts in the two destination regions comparable? The present research will examine some attempts made by tour operators to distinguish South Pacific and Queensland island destinations and the role of tour operators in the 'creation' of island resorts. These issues are taken up in Chapter 4 which examines comparative industry structure and Chapter 5 which examines comparative marketing.

Another important definitional issue in the present research concerns the terms 'tourism' and 'tourist'. Though tourists are popularly assumed to be synonymous with holidaymakers, the official definition propagated by the World Tourism Organisation (WTO) includes travel motivated by business, conference and visiting friends and relatives purposes. This all-encompassing definition is useful in urban destinations where a large proportion of travel activity is accounted for by business. It is less applicable for island resorts, where the almost exclusive motive for travel is holidaymaking and the resort environment is dedicated to satisfying the needs of this group of leisure travellers. The word 'tourist' is derived from the Latin *tournus* or circle (Boorstin 1964). The subjects of the present research are consumers participating in air-inclusive tours. The term 'circle' applies in that a circuit is followed starting from a generating region, proceeding through a destination and finally returning to the point of origin. Despite the fact that the word 'tourist' is technically a correct term to describe resort consumers, it is used sparingly in the present research. Depending on the appropriate context, the terms favoured in the present research are: 'leisure traveller', 'guest', 'consumer' and Krippendorf's preferred term, 'holidaymaker' (1987). Where the research applies to those travelling for both leisure and business purposes, the term 'visitor' is used.

Whereas the inadequacy of the term 'tourist' has been noted, 'tourism' is considered the most accurate term to describe the phenomenon as a whole. Distinctions between leisure, business and visiting friends and relatives (VFR) traffic are crucial, when we are examining consumers, separating out the impacts of business versus leisure travellers is often less critical when we are examining the supply side of tourism. The term 'tourism' describes the overall phenomenon adequately.

METHODOLOGY AND LIMITATIONS

In approach, this book is not aligned to any particular established academic discipline. It relies on an interdisciplinary approach, that has characterised much tourism research. The benefits of such an approach are clear – particularly in geographically confined areas such as the Whitsundays and the Mamanucas. To gain a true insight into such areas drawing from a wide range of perspectives is appropriate.

The methodology used in the research consisted of four main components. These were first an analysis of statistics and secondary sources, second, the conduct of consumer focus group discussions, third, a telephone survey of Australian travel agents and finally a series of semi-structured personal interviews with resort managers and other key decision-makers at the destinations, particularly destination marketeers. Some associated problems with the approach must, however, be recognised. No single researcher can be expected to cover all of the *relevant* sources where they are drawn from diverse origins including marketing, sociology and planning (Leiper 1981). The inevitable gaps are accepted by the author as a worthwhile sacrifice in the pursuit of a 'total' picture.

The statistics referred to and tabulated in the present study examine visitor profiles to the two destinations from the source markets and supply issues affecting the two island groups. As far as possible, the statistics are narrowed down to Melbourne and Sydney residents travelling specifically to island resorts in the Whitsundays and the Mamanucas by air. Where this level of precision is not available, the smallest available yardsticks are used (for example, holiday travel from NSW and/or Victoria and travel to the Whitsundays including the mainland and to Fiji as a whole). A broad range of secondary sources were evaluated covering all of the major themes covered in the research including tourism marketing, development, planning, environment and resort management in the two destination regions. Interpreting the available statistical data offers insights into the comparative performance of the two resort regions.

Statistical data is of good quality for the overall destinations of Queensland and Fiji. It is less comprehensive for the actual study areas of the Whitsunday and Mamanuca islands, though in the case of Queensland a limited range of data applied specifically to the Whitsunday islands were included in the Queensland Visitor Survey (QVS).

The Fiji Bureau of Statistics uses the area 'Mamanucas/Yasawas' as the relevant region for gathering accommodation statistics, thereby incorporating the Yasawa Islands which lie to the north of the Mamanucas. Fortunately from a statistical point of view, the Yasawas only offer two fairly small and very up-market resorts plus some capacity on cruise boats. This means that aggregate data can be applied to the Mamanucas alone with relative accuracy. Such data is supplemented by periodic studies on the Australian market conducted on behalf of the Fiji Ministry of Tourism by the New Zealand-based consultancy ASMAL, which separate data for the Mamanuca and the Yasawa Islands. In addition to collating appropriate government statistics, the author also gathered relevant research and reports from various government departments, private sector agencies and non-government organisations (NGOs) in both destinations.

Finding a suitable method to gather the views of actual and prospective leisure travellers to the Whitsundays and the Mamanucas proved an early and central issue in the overall research design. Any study that incorporates imagery is of limited value if it ignores consumer views. To rely entirely on surveys of intermediaries such as travel agents would have been insufficient without consumer reinforcement. To address this issue, eight consumer focus group discussions were staged with equal numbers of participants in Melbourne and Sydney (sixty-four participants in all). These two cities were selected because their combined population of over seven million constitutes almost 40 per cent of the total population of Australia (eighteen million), thereby offering the prospect of results that would have wide application and involve a manageable amount of travel for the author. Focus groups involve bringing together a number of consumers, usually between seven and nine per group, in a neutral setting to discuss perceptions of a particular product, group of products or concept. The approach was similar to that used in a recent US study in that discussion was held in purpose-built rooms belonging to an independent research firm (Milman 1993). Incentive fees were paid to participants and complimentary light refreshments were made available prior to and during the proceedings. The author acted as moderator and followed a pre-prepared flow chart to determine the timing of the various stages in the discussion process. A similar approach was used by Milman who states that 'following the "warm up" phase aimed to help the participants feel comfortable with the group process, each discussion was launched by introducing broad categories of vacation travel and incrementally narrowed to the topic of destination choice' (Milman 1993: 62). In the present research, participants were made aware that resorts were the focus for discussion after about five to ten minutes of introductory discussion. The sessions lasted one hour each on average. The Melbourne groups were videotaped. The Sydney groups were audiotaped because the room used was not equipped with a videotaping facility.

The focus discussion group method was selected for four major reasons. It is acknowledged by a number of authors as the best research approach for the identification of key issues and for evaluating consumer responses to different product concepts particularly when quantification of responses is not required or feasible (Peterson 1987; Gordon and Langmaid 1988). Since the present study has an interest in overall consumer perceptions (of the resort concept for example) as well as their responses to specific 'brands' (for example, 'Hayman Island' or 'the Mamanucas'), the method appeared to lend itself to the investigation. Second, the method is more cost effective than personal interview. Third, it allows the author to segment respondents into key demographic and experiential characteristics. Finally, a reputable market research company with extensive experience in running consumer focus groups on travel to Queensland, Brian Sweeney and Associates, offered the author logistical assistance with the groups. One Australian tourism academic has expressed scepticism over the rigour of the focus group method (P.L. Pearce 1988). He concluded that 'the focus group interview remains on parole'. He stressed the importance of reinforcing focus group studies with other sources and methods.

Certain limitations of the method must be acknowledged. Groups can be subject to the influence of an emerging collective mentality which may cause individuals to suppress their own views. It is also difficult to ensure that groups are genuinely representative of the demographic profiles that they are intended to represent. The quantification of responses is a questionable practice and often results in unrepresentative findings. For this reason no quantification was undertaken for the focus group research.

There were two possible alternatives to the use of the focus group method – interviewing in the source market using a different technique, or interviewing at the destination. Costs were a constraint. Undertaking consumer interviews is a costly exercise, especially where a relatively small proportion of the population exhibit the desired characteristics (i.e. they have visited the Mamanucas and/or the Whitsundays). The author could have undertaken a mail or telephone-based consumer survey in the source markets, but this course of action would have needed to be accompanied by a series of focus groups as well. Questionnaires can be more effective when the questions that they incorporate spring from ideas generated in a focus group setting. In the present research, the focus group discussions were used as a basis for the design of the travel agent survey. This approach is not perfect, since the two populations are dissimilar in many respects. The final selection of method was a compromise, brought about by the limited resources at the author's disposal.

An alternative method of data gathering was to ask consumers to keep diaries during their time in the resort. This method has been used previously by Pearce (1981). The author had concerns about the likely response rates and the accuracy of reporting, given that the research budget was not sufficient for an extended stay at the destination.

Groups were selected according to four key variables. Firstly equal numbers of groups (four each) were staged in Melbourne and in Sydney on the assumption that this geographical breakdown would help the author to identify any differences in consumers evident in the two areas. Also it was hoped that a reasonably representative segment of the Australian population could be examined through the selection of consumers in Australia's two major cities. Groups were also segmented into those travelling with families and those travelling as singles, or as couples without children. Age specifications were placed on both types of respondent, with older (over 45) and younger (under 25) consumers excluded from the study. Groups were also segmented according to their level of travel experience to the two relevant destinations. Three types of experience were examined, namely those who had visited both Queensland and Fiji, those who had visited one but not the other and finally those who had visited neither.

The 'line of enquiry' used to introduce the subject of resorts into the discussion involved six stages. These sought to elicit participant views as follows:

1 Mental associations prompted by the terms 'resort' and 'island resort'.
2 The information gathering and decision-making processes used in preparing for resort holidays were elicited.
3 The previous resort experiences of experienced travellers and the experiences anticipated by intending but inexperienced travellers.
4 Responses to various name prompts, both resort specific and regional.
5 Perceived differences between the two destination areas and the circumstances in which a visit to one would be preferred over a visit to the other.
6 Word association. Participants were asked to match certain descriptions to the two areas.

With Pearce's concerns about the inadequacies of focus groups as a stand-alone research method in mind, an extra quantitative dimension was added in the form of a telephone survey of 200 travel agents. As with the focus discussion groups, half of the respondents (200 total in this case) were to be in Melbourne and half in Sydney. The travel agents were selected randomly from the *Yellow Pages* telephone directory. Experienced telephone interviewers at Brian Sweeney and Associates were used to undertake the interviews. A random auditing system is used by the company to ensure that the interview work is done thoroughly and consistently. The questionnaire was drawn up by the author. Most of the questions were closed response with a small number of open-ended questions. To qualify, respondents were required to have booked clients to both Fiji and to Queensland. Seniority was not a requirement for respondents and the questionnaire could be answered by consultants, senior consultants or managers. Respondent seniority was recorded for cross-tabulation purposes. The telephone method was chosen because of its established reputation for a higher response rate achieved over equivalent mail surveys. Following a pilot study amounting to some thirteen completed interviews conducted by the author and some university research students, a number of questions were re-worded to streamline the finalised interview process. Experienced telephone interviewers from Brian Sweeney and Associates were used to conduct the actual interviews in preference to the author, or to the research students used for the pilot. The decision was based on the difficulties facing an inexperienced interviewer, particularly given the extended length of the interviews (20–30 minutes) and the reluctance of the busy target audience to respond to such surveys. The author had not previously undertaken telephone interviews and decided to work with the professional market research company to ensure the efficient and reliable conduct of the work, whilst retaining responsibility for questionnaire design. The experienced market researchers encountered a much lower refusal rate than the less experienced university researchers. This was an important consideration, since the total number of interviews could have fallen short of the original target of 200 had the initial refusal rate continued.

The finalised questionnaire started with some screening questions followed by eight main questions. A number of classificatory questions made up the remainder of the survey instrument. Questions investigated the level of awareness of the two island groups and of the individual island resorts within these groups, and whether the respondents had actually visited the islands. The strengths and weaknesses of the two groups as perceived by respondents was asked about. Next the travel agents were presented with a series of pre-determined traveller segments and asked to assess which group catered better to these segments. Three representative resorts were given from each of the two groups and respondents were asked to apply various descriptions to these resorts. The resorts selected were: Club Med, Hamilton Island and Daydream Island (in the case of the Whitsundays) and Plantation, Castaway and Mana Islands (in the case of the Mamanucas). Most of the descriptions were phrases used by participants in the focus discussion groups. Respondents were then asked about a number of industry issues, including any improvements that the tourism authorities in Queensland and in Fiji might bring about to improve the attractiveness of the resorts from a travel agent's point of view. The respondents were also asked about their agency's airline affiliation (if any).

The final component of the fieldwork consisted of semi-structured interviews with the managers of each of the resorts in the Whitsundays and the Mamanucas. All twenty-one resorts were covered with the exception of Plantation Island in the Mamanucas where the general manager refused to co-operate, despite numerous phone calls and faxes by the author. The author did gather some useful information about the resort from a recent senior manager in the resort and is personally familiar with the resort and its marketing activities. Extended interviews were undertaken with the resort general manager or proprietor. The only exception was Treasure Island in the Mamanucas where the general manager was compelled to leave the resort at short notice and cancel the interview. The author interviewed the assistant general manager instead. A number of interviews were also conducted with key industry and government figures at both destinations. The names of these people, their positions and the interview data are listed in the appendix.

Part I
THE ISLANDS

2

THE SETTING

The overview of the two island groups provided by this chapter gives readers a context and a point of reference for subsequent discussions. A geographical perspective is adopted to examine where the regions are located and the spatial relationship that links them with other regions and with the mainland in particular. The physical and human environments are also considered, with the needs and aspirations of the respective land-owners given particular attention. The landowning structures prevailing in the two regions are very different, tempting one to question the value of any comparison. In fact, the growing importance of native title as an issue in Australia has highlighted the potential benefits of learning from the South Pacific experience. South Pacific countries can also learn from Australia's experience of setting aside areas containing recognised heritage values. Long-term and sustainable tourism development of the two regions is critically dependent on high-quality cultural, environmental and heritage assets and this chapter devotes considerable attention to these issues.

THE TWO REGIONS COMPARED

The Mamanucas and the Whitsundays are both located in the south-west Pacific Ocean, far from the world's major tourism generating markets. Despite this, tourism is the key economic activity in both the Mamanucas and the Whitsundays. The Whitsunday group of islands is larger in size than the Mamanucas, consisting of seventy-four islands (see figure 2.1), as opposed to twenty-eight – more or less, depending on one's preferred definition (see figure 2.2). There are twelve resorts in the Mamanucas as opposed to eight in the Whitsundays (or nine, again depending on one's definition). Most of the Whitsunday islands are designated national park and have experienced minimal commercial development. The two groups are easily accessible from their respective mainlands, though the term 'mainland' has different connotations in the two instances. Though an island in the sense that it is unattached to any other landmass, Australia is a huge continent and hence almost the antithesis of an island. In contrast,

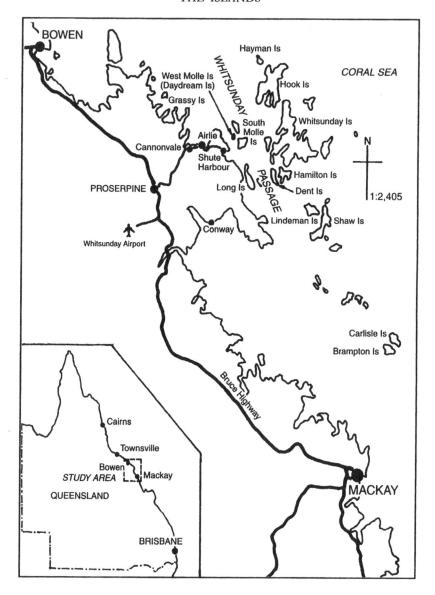

Figure 2.1 Whitsundays map
Source: B.J. Dalton (courtesy of the Department of History and Politics, James Cook University of North Queensland)

Fiji is described as 'the Fiji Islands' by the Fiji Visitors' Bureau, presumably to counteract the perception of a single landmass. Fiji consists of approximately 322 islands, of which the largest (often described as the 'mainland') is Viti Levu. We now examine the geography of the two island groups in turn.

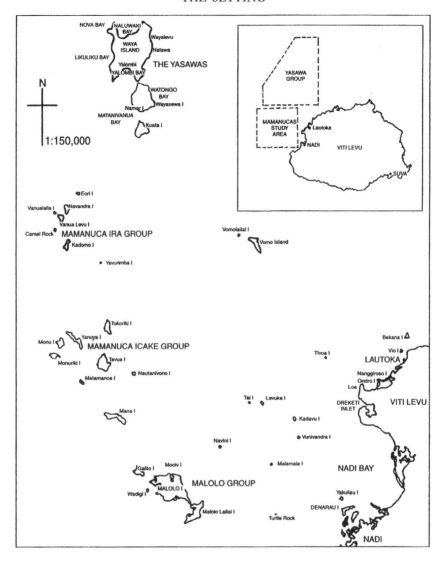

Figure 2.2 Mamanucas map

GEOGRAPHY

The Whitsundays

The Whitsundays are a chain of drowned mountains or continental islands situated off the North Queensland coast between Mackay and Bowen. They were previously described by Captain James Cook as the Cumberland Islands. According to Colfelt (1985) the generic term 'The Whitsundays' has emerged as the popular name for the whole of the Cumberland Islands. For tourism purposes, the group has sometimes been subdivided into the 'Central Group' of islands located around the Whitsunday Passage and the 'Southerly Islands' of Brampton and Carlisle. Though the operators of Brampton Island have periodically claimed inclusion within the Whitsundays Group, most definitions exclude the Southerly Islands. A recent management plan (GBRMPA 1993) attributes 208 square kilometres of national park to the Group and includes five sub-groups, namely the Gloucester, Whitsunday, Molle, Lindeman and Repulse groups. A number of reefs are also incorporated. The Whitsundays are within the area administered by the Great Barrier Reef Marine Park Authority (GBRMPA), though they are located some 40 kilometres from the Reef itself at the closest point (GBRMPA and the Queensland Department of Environment and Heritage 1993).

There are seventy-four islands in the Whitsunday group. The Whitsunday local government area (named Proserpine Shire until 1989) incorporates sections of the mainland including the towns of Airlie Beach, Proserpine, Cannonvale and Shute Harbour as well as the islands of the Whitsunday group. The population was recorded as 11,429 in 1991 with 14,830 projected for the year 2000 (Australian Bureau of Statistics 1992, QDHLG 1991). The adjoining local government areas of Mackay, Pioneer and Bowen are considered integral to the Whitsunday region for State Government planning purposes. The population for the four shires is projected to increase from 87,083 in 1991 to 117,629 by the year 2000 reflecting a growing employment base and population shifts in Australia away from inland and southern areas to coastal and northern locations. Apart from tourism and construction, the main regional industries are commercial fishing and sugar production. The increasing economic importance of tourism is evidenced by a comparison of agricultural production with hotel and motel earnings. In 1984–5 hotel and motel takings were 88 per cent of the gross value of agricultural production. In 1988–9 hotel and motel takings represented 129 per cent of the value of agricultural production (Office of the Co-ordinator General 1994).

The Whitsunday group includes small islands which are occupied substantially by a single resort (for example, Daydream Island), larger islands housing a single resort (South Molle and Lindeman islands, for example), one larger island featuring multiple resorts (Long Island which incorporates Palm Bay Hideaway and a Club Crocodile resort) and many with no resort development (the largest of the group, Whitsunday Island, has no resort development). Resorts cater for a variety of markets, including the exclusive (Hayman Island, for example) the 'middle' market (such as Daydream Island) and the 'budget' market (Hook Island). The Whitsundays have been described as constituting one of the four main island groupings off Queensland (King and Hyde 1989b) – or one of five according to Sunlover Holidays in its various brochures – and are probably the best known of the four (or five) groups. With the exception of the Southerly Islands (including Brampton) which are located within the Mackay Capricornia Zone, the Whitsundays form a part of the Central Zone of the Great Barrier Reef Marine Park. All of the resorts are classed as Marine National Park 'A' Category (a small part of Hook Island is 'B' Category). The zones are each the subject of a management plan. The Marine National Park 'A' Category aims to achieve 'protection of the resources of the park while allowing recreational activities and approved research' (Green and Lal 1991: 23).

The use of zoning is in itself a paradox. Unlike the situation in mainland national parks, marine parks such as the Great Barrier Reef allow for voluntary regulation rather than legally enforceable rules and permit commercial activities such as fishing and tourism. The Reef is regarded as possessing a 'right' to protection, but this should be balanced with the rights of users. Craik (1987a) has argued that the emphasis on voluntary codes has 'furthered the trend towards large-scale development of tourism through resorts, package deals and the like', since monitoring visitor activity is easier as part of an organised operation (1987a: 155). It is difficult to prove or disprove this claim, though it is clear that lavish investment has occurred without apparently being dissuaded by environmental management requirements. A more positive interpretation is that voluntarism may provide a more co-operative relationship amongst the key participating parties and persuade resort operators to take responsibility for the adjoining environment. The acid test is whether it will enable resorts to attract either more guests or else to persuade existing guests to spend more. This issue is critical to any examination of the extent to which guests involve themselves in the adjoining environment and are willing to do so.

The Mamanucas

The 300 sunlit, romantic islands of Fiji have won a world-wide reputation for their unspoilt South Seas charm and the friendliness of their people. Much of this reputation rests on a spectacular chain of palm and beach-fringed gems called the Mamanucas which lie in coral reef lagoons only a short distance from Nadi international airport.

(Siers no date given)

So begins Siers' eulogy to the Mamanucas published as a coffee-table book
in the late 1980s. Much that has been written on the islands uses this form
of promotional language. The present chapter explores the realities behind
the marketing and chapter 3 examines how the Mamanucas became one of
the two main resort regions in Fiji.

The Mamanucas are a group of islands located off the west coast of Fiji's
main island Viti Levu, close to the country's main international airport at
Nadi. Fiji itself consists of 322 islands of which one-third are inhabited. Of
the total land mass of 18,272 square kilometres, the two largest islands Viti
Levu (10,429 square kilometres) and Vanua Levu (5,556 square kilometres)
make up the bulk. The country was formerly a British colony and has been
independent since 1970. Tourism is the principal foreign exchange earner,
followed by sugar. Fiji is the major tourism destination in the South Pacific,
accounting for twice as many arrivals as its nearest rival, French Polynesia
and twelve times the number in Tonga. Its dominance is relative, however,
with northern Pacific destinations attracting larger numbers – Hawaii
attracts twelve times as many and Guam twice as many as Fiji. Australia
accounts for almost one-third of all visitor nights in Fiji, making it the
dominant market source (Yacoumis 1992).

The Mamanuca islands range from 7–40 kilometres off the Viti Levu
coast and most of the inner islands can be reached by boat departing from
Denarau Beach (near Nadi). The more northerly islands can be reached by
boat from Fiji's second city, Lautoka (population about 40,000). An airstrip
is located on Malolo Lailai Island, home of two of the Mamanuca resorts –
Musket Cove Resort and Plantation Island Resort. Sunflower Airlines and
Fiji Air operate the ten-minute flight from Nadi International Airport six
times daily. Castaway Island can also be reached by seaplane and Tokoriki
and Matamanoa by helicopter. The facilities of the Mamanuca resorts are
smaller in scale than their Whitsunday counterparts with the largest, Mana
Island, having a capacity of 132 units. Most are targeted at the middle rather
than the upper end of the market, though the recent opening of the tiny
Sheraton Vomo Resort, with a total capacity of thirty suites, has added an
up-market alternative to the range, only accessible to visitors by helicopter.
Most tourist accommodation in the Mamanucas consists of *bures*, or 'traditional' Fijian dwellings, typically thatched, single storey and detached.

According to Derrick (1951: 221–3), the Mamanucas incorporate two
distinct geographical areas, namely Mamanuca-i-Ra (the leeward Mamanucas) and Mamanuca-i-Cake (the windward Mamanucas). These two areas
are made up of the islands listed in the first two columns of table 2.1.

The island of Vomo (which is now occupied by a Sheraton resort) and
Vomolailai (floodlit for the pleasure of dinner guests at Sheraton's On the
Rocks restaurant) are described by Derrick as 'not properly included in the
Mamanuca group'. They are, however, regarded as an integral part of the
Mamanucas by Cabaniuk (1992) and it is the latter point of view which is

Table 2.1 The sub-regions of the Mamanuca Islands: component islands

Mamanuca-i-Ra	Mamanuca-i-Cake	Malolo Group
Yavuriba	Tavua	Malolo
Kadomo	Yanuya	Malolo Lailai
Vanualevu	Monu	Qalito
Navadra	Monuriki	Malamala
Eori	Tokoriki	Navini
	Mana	Elevuka
	Matamanoa	Kadavulailai
	Nautanivono	Islets, e.g. Mociu, Wadigi and Vunivadra

Source: Derrick, 1951

followed in the present research (see also figure 2.2). The explanation for including Vomo in the present study is similar to the one explaining the inclusion of Brampton as part of the Whitsundays. Because both islands are on the margins of their respective regions, their relations with the respective groupings can provide some insights into interactions between the various participant resorts. Brampton has shown greater enthusiasm to secure membership of the Whitsundays than Vomo has in the case of the Mamanucas. The challenge of selecting what precisely should be included under the regional tourism label is a problem worldwide and the present study explores some of the key issues associated with this.

The islands of Malolo, Malolo Lailai and Qalito ('Castaway Island') may be considered as the 'heart' of the Mamanucas for the purposes of the present study, because of their longevity as resorts and their excellent transport links, though Derrick declined to classify the core islands of the 'Mamanucas' tourism industry as Mamanucas at all and used the label 'Malolo Group' to describe the islands listed in the third column of table 2.1. For administrative and planning purposes the Fiji Government classifies the Mamanucas into the three groups listed above, and supplements these with the islands of Namotu and Tavarua in the South. Three further small islands located close to the coast of Viti Levu, namely Tivoa, Yakuilau and Bekana are included in Cabanuik's study of the Mamanucas. Although geographically distinct from the rest, they have important links with the tourism industry, with Fijian cultural traditions and (in the case of Yakuilau), with ancestral legends. The total number of islands is twenty-eight, excluding the various islets (three of them listed under the Malolo Group). The present study focuses on those of the twenty-eight islands which have resorts, but the remaining islands are studied as a valuable backdrop to activity on the resort islands themselves.

The Fiji Bureau of Statistics groups the Mamanucas together with the more northerly Yasawa islands in its recording of accommodation trends. The Mamanucas are considered part of the 'Western Division' of Fiji. The more northerly of the islands are administered by the Ba Province and the more southerly ones by the Nadroga and Navosa Provinces. The former province extends eastwards from Lautoka and the latter southwards from Nadi. The administrative divide might be expected to lead to problems in achieving a genuinely coherent approach to regional tourism. The total population of the Mamanucas/Yasawas area is only 7,000 out of a total for Fiji of approximately 750,000, though it attracts over one-fifth of all visitors to Fiji and has a higher ratio of visitors to residents relative to other regions of Fiji.

Having now examined the location and physical characteristics of the two groups, we now turn to the issues of land tenure and management. An understanding of this background is an essential prelude to comparing the nature of tourism in the two destinations, with particular respect to indigenous involvement in tourism.

LAND TENURE AND MANAGEMENT

The Whitsundays

Most of the land mass in the Whitsunday islands is classed as Crown Land. This is in contrast with the situation prevailing in the Mamanucas, where most land is classed as Native Title. The contrast exemplifies the fact that in Australia, the Government exerts the greatest influence over most land coinciding with or adjoining areas of particular natural significance. In Fiji, the Government has less direct authority since it must always contend with the interests and rights of the native landowners. Prior to the first influx of tourism into the Whitsundays, most of the land was leased for grazing, though much was not economically viable. Some timber gathering was also undertaken. Early promotional efforts by the then Queensland Government Intelligence and Tourist Bureau (QGITB) 'while encouraging tourism, were largely orientated towards attracting agricultural and pastoral settlement' (Barr 1990: 8). This was comparable to the early activities of the Fiji Publicity Board which sought to attract migrants rather than tourists. The dominant form of land use remained pastoral until the 1930s when rudimentary tourism resorts were commenced. Barr reports (1990: 9) that 'the visitors were largely self-sufficient, bringing their own provisions and sometimes erecting temporary camps on the islands'. Most visitors were from the local area. A typical example of catering for visitors on St Bees (now Brampton) was putting a few mattresses down in the shearing sheds!

The earliest developments are symptomatic of the cyclical nature of tourism. The potential of the area for tourism was first realised by what would today be called 'ecotourists' or 'educational tourists'. A tour to Lindeman during Christmas 1928 involved over 100 scientists, teachers and other parties interested in undertaking a survey of fauna and marine life in the locality. A whole series of tours was undertaken by groups known as 'expeditionalists'. It is ironic that ecotourism is regarded as a contemporary activity when it is better described as a contemporary adaption of a long-established activity. The early history of tourism is clear evidence that the attractions of the Whitsundays are sufficient to draw ecotourists, or at least tourists with an interest in the environment.

The separate ambitions of scientists and of tourism developers were influential in the conversion of the earlier grazing leases into national parks and into tourism leases respectively. Hundloe, Neumann and Halliburton (1989) regard the watershed for management of the region as the formation of the GBRMPA in 1975. Prior to that the Reef area was viewed by many as an open access common property 'because as a general rule, management regimes were poorly defined and few restrictions imposed' (Hundloe *et al*: 9). Before 1975, management of the in-shore territorial seas had been in practice, if not constitutionally, under Queensland government jurisdiction. This confusion was a symptom of the dilemmas of a Federal political system. The authors clearly attributed the establishment of the new management authority to resolution of the dilemma. The presence of a strong regulatory authority is not mirrored in the Mamanucas.

To date there has been minimal indigenous involvement in Whitsunday tourism, symptomatic of the Queensland Government's overall attitude to Aborigines. In 1994, *The Australian* newspaper reported that 'more Aborigines should be appointed to wilderness management boards such as the Great Barrier Reef and the Queensland Wet Tropics, the Federal Government said yesterday as it handed over a fourth national park to Aborigines' (Carruthers and Cant 1994). The implication was that the (at that time Labor Party) Commonwealth Government was dissatisfied at the slow pace of reform being implemented by Queensland and other State Governments to ensure indigenous participation in the key decision-making bodies.

Blainey, in his *The Triumph of the Nomads* (1976: 225) asserted that prior to European settlement 'if we specify the main ingredients of a good standard of living as food, health, shelter and warmth, the Aboriginal was probably as well off as the European in 1800'. According to Tindale (1974) and Barker (1992), Aborigines of the Gia and Ngaro clans resided in the Whitsunday region prior to Cook's arrival. Barker has estimated that 300 Aborigines lived on the islands. European settlement led to a catastrophic drop in the indigenous population. The historian Ross Fitzgerald (1982) has estimated that the Queensland Aboriginal population dropped from 100,000 in 1840 to 15,000 in 1900. In the twenty years following initial European settlement of the Bowen, Proserpine and Mackay districts in the 1860s virtually the entire Whitsunday population was wiped out, though there are some descendants of the original inhabitants of the region resident in the mainland Whitsundays (Office of the Co-ordinator General 1994). Through the nineteenth century the colonial administration regarded Aborigines as a 'doomed race' unsuited to work in European-style enterprises. One exception was George Brigman's Mackay Reserve where 200 Aborigines were involved in the cutting and harvesting of sugar cane in the 1870s. Any prospect of the maintenance of a stable Aboriginal population ended in 1878 when the State Parliament cancelled the provision of funds to the project. Over half of the 4,000 acres accounted for by the Reserve were quickly appropriated by incoming settlers.

Historically, Queensland has been regarded as one of the more oppressive governments towards its Aboriginal population. An increasing liberalisation towards Aboriginal affairs at Commonwealth level was exemplified by passage of the *Racial Discrimination Act 1975* (Cwlth) by the Whitlam Labor Government and passage of the *Aboriginal Land Rights (Northern Territory) Act 1976* by the previously reluctant Fraser Liberal Government. More recently the *Aboriginal and Torres Strait Islander Commission (ATSIC) Act 1989* (Cwlth) sought to 'maximise participation by Aboriginal and Torres Strait Islander persons in the formulation and implementation of government policies that affect them' and promoting 'self-management and self-sufficiency' (quoted in Bergin 1993). In contrast, the Queensland Government continued to argue for the legitimacy of perpetual leases in traditional Aboriginal lands whilst opposing any lease transferral to Aboriginal groups or even to individuals on behalf of such groups. Aboriginal Land Councils were not recognised by the State Government. The various amendments to the *Forestry Act 1959* (Qld), enacted between 1976 and 1987, pitted Aborigines against conservationists by effectively excluding Aboriginal occupation of state parks and forests 'through its bar on such activities as hunting, gathering, fishing, building huts and its strict controls on lighting fires' (Fitzgerald 1984: 539).

A remark made in 1981 by Queensland Minister for Aboriginal and Torres Strait Islander Affairs, Ken Tomkins, highlights the huge difference between indigenous land management in Fiji and in Queensland. Tomkins stated that 'I'm not satisfied at this point of time that Aborigines can handle mortgage documents' (quoted by Fitzgerald 1984: 543). In contrast, the Fiji Government made an early decision to ensure that Fijians employed within the Native Land Trust Board, which handles all commercial activity relating to native land, are well versed in land management issues. The Queensland National Party Government was voted from office in December 1989 and the incumbent Goss Labor Government adopted a more inclusive approach to the State's Aborigines. Even the Goss Labor Government was slow to act on Aboriginal land ownership and its legislation, the *Aboriginal Land Act and Torres Strait Island Land Act 1991* (Qld) made no allowance for the hoped-for land acquisition fund (Kennedy 1995). The long history of exclusion from any development highlights how totally divorced tourism developments such as those in the Whitsundays have been from the indigenous population. Exclusion is the predominant historical legacy of resort development. Not that the Whitsundays are atypical of Australia. According to the *Draft Commonwealth Aboriginal and Torres Strait Islander Tourism Strategy* (ATSIC 1994) there were only 2,500 indigenous Australians employed in the tourism industry in 1991 representing a mere 0.7 per cent of all people employed in the industry, though it proposes to expand this number to 11,000 by the year 2000 or 1.9 per cent of all tourism employees.

Unlike many accounts of Australia's tourism development, Leiper's interdisciplinary study of Australian tourism is unusual in having highlighted the existence of Aboriginal tourism in Australia prior to European settlement (1980). He argued that Aboriginals had leisure time, economic independence and motivation and that visits they made to kinsmen and religious places constituted an early form of tourism. The recognition of such continuity is likely to grow in importance as contemporary indigenous tourism emerges as a significant force.

The Mamanucas

The complex landowning structure in Fiji impacts upon relations between the resorts and the community. Indigenous Fijians play the predominant role in provincial administration because the former British colonial authorities chose to build an administrative system based upon the existing Fijian chiefly structure. Nayacakalou (1975) has described the Fijian administration as 'a system of local government for Fijians'. Administration takes place at three levels:

- **The Village**   Each village has a village council headed by a *Turaga ni Koro*.
- **The Tikina**   This includes a number of villages and is headed by an Assistant *Roko*.
- **The Province**   This is under a Council headed by a *Roko*.

Adding to the complexity, different labels and groupings apply to issues of landownership. Fishing and land rights, for example are held at different levels. In the case of fishing rights, the *Yavusa* (incorporating people from a number of villages) is the relevant unit, whereas land rights are held at the *Mataqali* level. The *Mataqali* is constituted of kinsmen who are descendants (determined through patrilineal inheritance) of the founder of the *Yavusa* (Ravuvu 1987). *Mataqali* members are often spread over a variety of locations and there are usually a number of *Mataqali* in each *Yavusa*.

Most of the Mamanuca island group lies within the Province of Nadroga and Navosa and within the *Tikina* of Malolo. Malolo is considered as a

Vanua made up of various *Yavusa*. These are listed by the Native Land Trust Board (NLTB)'s Classification of Communal Units – Province of Nadroga. The head of the *Vanua* holds the title of *Tui Lawa* and traditionally resides in Solevu Village on Malolo Island. The *Yavusa* of Malolo are shown in table 2.2.

Table 2.2 is indicative of the complicated nature of landowning. Fishing rights are also determined on the basis of such classifications. Residents of the village of Solevu on Malolo Island may be members of one of two Yavusas – the Lawa or the Taubere. The Yavusa Lawa also has members in the adjoining village of Yaro (also on Malolo Island). The village of Yanuya on Vunaivilevu has its own Yavusa (Yavusa Yanuya). It may be readily appreciated that this complex landowing structure can create social and cultural differences which may result in problems for resort management.

Since some of the Mamanuca islands are not in the Province of Nadroga and Navosa, they lie outside the influence of the Tui Lawa and the Vanua of Malolo. The following islands fall under the jurisdiction of the Tui Vuda, Ratu Josaia Tavaiqia who resides in Viseisei Village: Tai, Levuka, Navadra, Vomo and Kadavulailai. The different dynamics and personalities involved make it difficult to generalise about community linkages across all of the Mamanuca island resorts. The following islands fall under the jurisdiction of the Tui Nadi, Ratu Josua Dawai, who is a resident at Narewa Village: Malamala and Yakuilau. Finally the island of Bekana, off the coast of Lautoka is under the jurisdiction of the Taukei Vidilo who resides at Namoli Village. The role of the relevant chief is very important for establishing a working relationship between the villages and the resorts, since it provides a clearly defined hierarchy for the implementation of decisions. In the present study, the close integration of the community and the resorts is noted. It should also be noted that the complexity of administrative and social structures could merit a study in its own right.

Only one island in the Mamanuca group is classed as freehold, namely Malololailai. The remainder are classified as native reserve. Any use other than for traditional Fijian purposes (tourism is certainly not classed as 'traditional') requires consent from a majority of the relevant landowners and from the NLTB. This is in stark contrast with the system of land-

Table 2.2 List of Yavusa within Malolo (villages listed in parentheses)

Yavusa Lawa	Lawa (Solevu, Malolo)
Yavusa Taubere	Malololailai (Solevu, Malolo)
Yavusa Yanuya	Vunaivilevu (Yanuya)
Yavusa Lawa	Lawa (Yaro, Malolo)
Yavusa Leweimotu	Motu (Tavua)

Source: NLTB 1980

holding in the Whitsundays where commercial criteria predominate, providing that environmental requirements are adhered to. The Fiji approach to tourism development is more consciously 'holistic' than its Whitsunday equivalent.

ENVIRONMENT AND CULTURAL HERITAGE

The Whitsundays

The respective roles of the Commonwealth and State governments have been the subject of heated debate during the 1980s and 1990s. The Commonwealth has increasingly invoked its external affairs powers to bring about Australian adherence to international treaties and agreements and to bring about World Heritage nominations and listings. States such as Queensland fought such moves, regarding them as an interference with States' rights which they argued embodied the true spirit of the Australian Constitution. Environmental issues were the main battleground for such confrontations. With environmentalism rising throughout the world, successive Labor governments in Canberra sought to further develop the environmental legislation put in place by the Whitlam government in the early 1970s. With increasing votes to be gathered from pursuing a manifestly environmentalist line and a philosophical commitment to reform, Canberra decided to expand the scale of environmental protection in Queensland. In Brisbane, the Bjelke-Peterson-led Queensland Nationalist government was driven by a different coalition of interests, principally the electoral gerrymander which gave disproportionate political clout to conservative rural voters. The State Government was hostile to what it dubbed as the 'socialists' in Canberra and sought to exempt Queensland from what it portrayed as excessive bureaucratic interference. The policy nevertheless allowed for state intervention in tourism and other development projects. Rural interests were protected, whilst developers were often given the green light with little emphasis on environmental repercussions. Bjelke-Peterson's enthusiasm for Keith Williams' Hamilton Island development was typical of the philosophy of the times.

Enter the Commonwealth Government with its attempt to 'lock up' major tracts of Queensland through World Heritage listing. The Queensland Government bitterly opposed these moves, seeing them as a deliberate erosion of State rights at the behest of 'greenies' and 'socialists' down south. The contest took on farcical proportions when the Queensland and Commonwealth governments each dispatched delegations to argue their respective cases to the relevant agency in Brazil. In the event, the Queensland Government lost the argument and World Heritage listing was approved. The Nationalists also lost the electoral battle. The party was defeated at the subsequent State election and the Federal election was won

37

by Labor, greatly assisted by the environmental vote. In contrast to its National predecessor, the incumbent Queensland Goss Labor Government has elevated environmental and heritage issues to centre stage (critics claim it has been too timid) and has worked closely with the Commonwealth Government on joint legislation.

The GBRMPA was established through the *Great Barrier Reef Marine Park Authority Act 1975* (Cwlth). An important strategic objective of GBRMPA as outlined in its twenty-five-year strategy is the enhancement of Aboriginal involvement in park management (GBRMPA and the Queensland Department of Environment and Heritage 1993). This involvement could have significant repercussions on the Whitsundays. A report by Bergin (1993) for the GBRMPA, *Aboriginal and Torres Strait Islander Interests in the Great Barrier Reef Marine Park,* proposed an increase in Aboriginal participation. Whilst commending the GBRMPA on the professionalism of its management, the report observed the lack of Aboriginal involvement as a deficiency, particularly relative to the World Heritage listed Kakadu and Uluru National Parks where Aboriginal involvement includes management, operational and interpretative roles. The report pointed to the significant Aboriginal populations in areas adjoining the Reef at Thursday Island, Palm Island and Yarrabah.

The fact that Aboriginal issues will be increasingly prominent in the Whitsundays was flagged by a recent native title claim covering land between Bowen and Proserpine, including the Whitsunday islands, by a group of 172 direct descendants of the Dagaman people (*Sunday Age* 1994: 2). Bergin (1993) had previously pointed to the likelihood that such claims would arise. Whilst the absence of a permanent Aboriginal population in the Whitsunday islands means that any indigenous involvement in Park management is likely to be low key (the Dagaman proposition was described as an 'ambit claim' by Premier Goss and seems to have centred around a cattle station forty kilometres west of Proserpine). Though the type of community liaison proposed by Bergin as appropriate for areas with permanent populations is less likely in the Whitsundays, there is undoubtedly a prospect of greater Aboriginal involvement as park rangers, as participants in research projects to identify significant sites and in overall park management (Bergin (1993) points out that the Boards of Management in Kakadu and in Uluru are empowered to approve or disapprove the relevant management plan and have a majority of Aboriginals). Given that Club Med relies on National Parks and Wildlife Service rangers to show resort guests around Lindeman Island, it is likely that visitors will come into contact with Aborigines occupying such roles in future.

The Commonwealth Government's *National Ecotourism Strategy* was launched in 1994. Indicative of the close alignment of ideas between the State and Commonwealth Governments, the Queensland Government has developed its own ecotourism strategy, in draft form at the time of writing

(DTSR 1994). The Commonwealth Government's commitment of A$10 million over the period 1994–8 to help implement the National Ecotourism Strategy may also have prompted the State Government to act. The Commonwealth strategy uses the following definition of ecotourism: 'sustainable tourism focused on natural (and associated cultural) attractions which is both educational for visitors and beneficial for destination host communities'.

A number of issues referred to in the Draft Ecotourism Strategy are of relevance to the Whitsundays. Firstly the State has been divided into a series of bio-geographic regions for the development of ecotourism. The Whitsundays are incorporated within the 'Central Mackay Coast' region, which stretches from Mackay to Bowen inclusive. This is wider than the normal planning boundaries for the Whitsundays. The Whitsunday Islands and the Great Barrier Reef World Heritage Area are listed as sites of national and international significance with scope for ecotourism development. Four bio-geographic regions are singled out as having 'outstanding potential', namely the Great Barrier Reef and the Central Mackay Coast. The strategy determines that settings offering potential for ecotourism are a 'combination of physical, biological, social and managerial conditions that give value to a place' (DTSR 1994: 16).

The Draft Strategy lists a number of activities which it would class as ecotourism. A number of these can be readily applied to the Whitsundays, namely:

1 'A boat trip to the Great Barrier Reef which presents and interprets marine and reef ecosystems';
2 'A forest walk with an Aboriginal guide explaining the Aboriginal culture of the local area, traditional food and medicine, resources, native plants and animals';
3 'Bird watching in a natural environment';
4 'Participating in studies with scientific or environmental research groups';
5 'A guided bushwalk through a local area of remnant forest'; and
6 'Staying at accommodation designed to integrate and educate visitors in the natural environment and minimise impacts through sewerage, waste disposal, power generation use, landscaping, design and appropriate hardening of tracks' (1994: 12).

A number of recommendations in the Draft Strategy could impact directly on the management of Whitsunday island resorts, including the preparation of 'standards and guidelines to minimise generation of waste products, promote environmentally-friendly disposal of waste products and maintain aesthetic values' (1994: 17).

The Draft Strategy considers the involvement of local people in tourism as an integral part of ecotourism. It emphasises the potential (and as yet largely unexploited) opportunity for Aborigines. 'The relationship of indigenous people with many of Queensland's protected areas and other natural resource areas should be incorporated into planning and management of natural resources' (1994: 19). Whilst many of the principles espoused in the Draft Strategy are superficially appealing for the Whitsunday islands, a number of dilemmas are also evident. One may argue that the strongest competition for the Whitsundays is the Gold Coast and there is some risk that pursuing ecotourism at the expense of the existing mass market could result in a loss of competitiveness relative to the primary competition. On the other hand, another key competitor, Far North Queensland has given increasing emphasis to rainforest and other ecotourism activity. The Whitsundays Visitor Bureau (WVB) will need to position the region with great care relative to such different competitors.

A *Draft Management Plan for the Whitsunday National and Marine Parks* was prepared by the GBRMPA and the Queensland Department of Environment and Heritage (QDEH), proposing strategies to 'protect the natural, cultural and heritage values while allowing the public to continue to use and enjoy the wide range of recreational and other activities that are permitted' (1993: i). The foreword to the document states that 'The Draft Plan places a strong emphasis on co-operation with island resorts'. The Plan was designed to complement the Queensland Government's *Coastal Protection Strategy*. It remains in draft form only as at mid-1996, indicative in part of a likely failure to achieve consensus amongst the parties consulted and also because of administrative uncertainties associated with the change of government.

The Whitsundays have emerged as a clearly defined tourism region with a co-operative relationship between the various operators and accommodation houses in the pursuit of regional objectives. Tourism is acknowledged as the key economic activity within the islands. The region is now characterised by relatively large resorts operated by large hotel management companies. The 'traditional' informal-style resorts are now marginalised and are seen as representing an 'alternative' style of holiday for a small group of *aficionados*, now in the minority. Though the larger resorts still preach a philosophy of informality, the reality is a higher level of institutionalisation than existed previously, the predominance of hotel-style resort construction (as opposed to the self-contained units of the Mamanucas) and a greater adherence to an 'international style' with increasing numbers of overseas visitors. The other notable aspect of the Whitsundays has been the disappearance of the Aboriginal population which was so actively depicted by early authors writing about the Queensland islands. Though Aborigines and more particularly Torres Strait Islanders played a part in Whitsunday tourism in the interwar years acting as entertainers and musicians, they are nowhere to be seen in current imagery or activities of the island resorts. The Australian High Court's Mabo decision on Aboriginal land rights and the appointment of an Aboriginal Tourism Unit within the

QTTC are indications that the cultural heritage issue needs to be taken more seriously by Whitsunday tourism operators.

In contrast to the anti-environmental and government planning rhetoric of the earlier National Party Government, the Whitsundays have recently been the subject of a plethora of planning and environmental studies, many of them instigated by the State Government. The Draft Whitsunday Tourism Strategy, the Draft Management Plan for the Whitsunday National and Marine Parks and the Queensland Ecotourism Strategy (to name only three) are symptomatic of a convergence of marketing, environment and planning into a coherent whole, though the fact that the first two of these documents were still in draft form only as of mid-1996, is an indication of delays in bringing about a coherent plan-based approach. In light of the diversity of the Whitsundays with its contrasting islands, coastline and hinterland, some formal mechanism was clearly required to draw together the various relevant parties. One can readily appreciate the likelihood that internecine quarrels would emerge over a contentious issue such as tourism where different organisations driven by very different philosophies (for example, environmental, entrepreneurial, bureaucratic) and representing contrasting constituencies came into contact with one another.

A number of the relevant tourism plans and strategies are still in draft form at the time of writing and changes may be made to the proposals of these documents following the consultation processes entailed for each document. Implementation of the various proposals will face a number of hurdles. The diversity of the region is an advantage in marketing terms, but what if accommodation and other facilities fail to deliver the type of service standards expected by visitors? Is there not a danger that the Whitsundays area as a whole could be sullied by substandard product? Will a Whitsunday tourism style emerge? The *Draft Whitsunday Tourism Strategy* (Office of the Co-ordinator General 1994) expressed a desire for the application of regional design criteria to new resort and building development. It is of course difficult to persuade private developers to build in a certain way if they are not coerced or encouraged with hard-to-justify taxpayer subsidies. The question remains as to whether a distinctive Whitsunday style will prove to be more than rhetoric. Even allowing for such difficulties, the strategic and management planning approach will allow problem areas to be addressed in a systematic fashion. It was notable that no resort island manager interviewed by the author actively criticised the involvement of mainland interests within the Whitsunday Visitors Bureau (WVB). The Whitsundays are well placed to take advantage of prospective tourism growth with the necessary structures in place to resolve potential disputes along the way, though it should be noted that the area remains dependent on discounted air package initiatives and that the small population of its hinterland limits the range of activities and shopping available to holidaymakers.

The Mamanucas

After independence in 1970, the Fiji Government continued the practice of earlier colonial administrations in preparing a series of five-year development plans (DP's), before moving to a more deregulatory approach in the late 1980s with the last DP9 concluded in 1990. Ironically, it has been during this period of deregulation that environmental issues have been elevated in importance. Fiji is now a signatory to no less than nineteen international environmental conventions, but few if any are thoroughly implemented.

In 1989 the Fiji Government instigated a two-year project to prepare a National Environment Strategy for Fiji. The result of the investigation was the *State of the Environment Report* (IUCN 1992), a sort of 'stocktake of environmental quality, natural resources and their use, environmental policy and law and environmental administration'. The Report identified 140 sites of national significance, of which eight were in the Mamanucas area, as shown in table 2.3.

A major limitation of the Report is that it does not undertake any further analysis of the suitability of these sites for tourism development. Though the document provides some valuable baseline data, the items listed vary enormously in their ability to sustain larger numbers of visitors and also in their potential tourism appeal. It is unlikely that the attractions will be adopted by tour operators in Australia, unless they are clearly identified and assessed regarding their suitability for tourism. In contrast, studies conducted by the GBRMPA on the area under its management (including the Whitsundays) have been cognisant of the relationship between tourism operators and site-based attractions. The failure to address the linkages between potential heritage attractions and tourism in Fiji is symptomatic of

Table 2.3 Preliminary register of sites of national significance

Location	Type of attraction
Whole Mamanuca Group	Coastal/Marine Ecosystem, Recreation
Nadi Bay Reefs	Reefs, Recreation
Malamala Island (Malolo Group)	Marine Ecosystem
Vanualevu Island (Mamanuca-i-Ra)	Geological Site-Rock Type
Kadomo Island (Mamanuca-i-Ra)	Shearwater Nesting Colony
Monuriki Island (Mamanuca-i-Cake)	Iguana Habitat, Seabird Nesting Colony, Recreation
Vomosewa (Other Island Category)	Flying Fox Camp / Island Vegetation
Vunivadra Island (Other Island Category)	Seabird Nesting Colony

Source: IUCN 1992

42

the less sophisticated approach to tourism and the environment relative to the Whitsundays.

The challenge for environmental management in Fiji is summed up by the IUCN (1992) as follows:

> At least 25 acts have some important role in what is today perceived as environmental management, and they are administered by at least 14 different ministries, statutory bodies or other agencies. Most of the laws are old and ineffective in a modern environmental management context or suffer from lack of enforcement through inadequate staffing, lack of technical resources and funding or through administrative failures.
>
> (1992: 16)

Nor is Fiji's developing country status regarded as proper justification. As the report states (1992: 16), 'Although neighbouring Pacific nations have internationally recognised national parks, Fiji has none'. The report also points out that in Fiji 'there is no National Register of archaeological or cultural sites of historic interest' (1992: 143). These criticisms highlight the problem that the Mamanucas must confront, if they are to diversify their tourism appeal and keep up to date with international trends linking tourism and the environment. They highlight key differences between developed and developing countries that need to be acknowledged, before the development of so-called 'ecotourism' is undertaken.

The Report also expresses the view that although 'Fiji still retains a largely unspoilt environment' (1992: 146), 'pollution problems are evident' and that 'waste disposal is a national dilemma; the location and management of every municipal tip in Fiji indicates a total disregard for internationally acceptable standards' (1992: 147). The Nadi municipal waste dump, destination of much of the refuse from the Mamanucas resorts, is located by mangroves at the shore and is privately managed. It is one hectare in size and undertakes compaction and levelling. No pollution has been reported, though sea pollution is suspected. The watertable is shallow and the dump is full. The limited disposal facilities are a constraint to any sustainable expansion of tourism capacity. Issues of public health are vital, irrespective of whether one is developing 'sun, sea and sand' tourism or ecotourism.

With the exception of some glass products, there is no significant recycling of waste in Fiji. 'The major reason is the insufficient quantity and the dispersed distribution of the waste which makes it uneconomical to invest in a recycling plant.' The Report recommends that 'government or municipal authorities should offer economic incentives to attract investment' (1992: 112). The Report identifies that wastes from the tourism industry are:

> primarily sewage from the guests, restaurants and staff. Sewage treatment systems at the main hotels are usually quite sophisticated and

well-run, with sewage often undergoing primary and secondary treatment. The quality of effluent is monitored by some of the larger hotels, but there is no monitoring of beach or lagoonal water quality in these areas. Smaller tourism facilities have septic tank systems, which are usually adequate, but expansion of such facilities has in some cases involved an undesirable density of such tanks.

(1992: 103)

As yet, the issue of environmental responsibility is probably not a key determining factor for consumers when choosing between alternative destinations. In due course, however, it could emerge as a challenge for destinations such as the Mamanucas.

When traditional practices and modern technology interact, non-sustainable practices may be the result and the Report points to inefficient water management within the Mamanucas, resulting in the need to barge in water supplies. Comparable environmental audits in the Whitsundays area have not encountered such problems.

Fiji's Western Division experiences less, but still adequate, rainfall relative to the Central Division (the east side of Viti Levu). Between two and four cyclones pass within 70 kilometres per decade causing severe damage. Three have affected the Mamanucas area since 1972, namely Cyclones Bebe (1972), Oscar (1983) and Nigel (1985). The 3–4 metre swell in Nadi Bay prompted by Cyclone Oscar caused major resort damage. Annual average rainfall is approximately 1,400 mm, sufficient for 56 per cent of the community's annual needs if each household in the village were to upgrade its method of rain collection. If properly maintained the Kawabu source could supply the remaining water needs for the community. Water supply is a crucial environmental issue in tropical island resorts if a sustainable tourism development is to be achieved.

Should a system of national parks and protected areas be established, traditional landowners 'will rightly require an income equal to or greater than any other potential landuse' (1992: 86). Landowners want a continued livelihood, based on an integrated approach to development activities – 'compensation' may not be a sufficient response.

Traditional methods are not always positive for the environment and the Report acknowledges that some environmental problems have been ignored by government out of a reluctance to confront landowners. Such negatives include 'traditional rights (for example, fishing reefs), foreshore reclamation, rivers and streams, agricultural practices and land use etc.' (1992: 148). This observation is a reminder that relations between the resorts and the environment involve parties other than direct tourism interests. This has not been a problem in the Whitsundays because of the absence of a resident population.

According to the Report, the Mamanucas and other marine environments are less badly affected by environmental degradation than other parts of the country, particularly those that have experienced deforestation (1992: 132). Inconsistences do exist, however. Most species of birds in Fiji are afforded protection, though reptiles are ignored. Whilst turtles are afforded some protection under the *Fisheries Regulations 1965* (Fiji), the internationally renowned crested iguana (reputedly found in the Mamanucas) has no legal protection. In Fiji only one National Monument has ever been declared – Wasavulu near Labasa. This National Archaeological Monument is described as 'a site of unknown significance or original function' which is 'sadly neglected' and under uncertain management. The dilemma of such sites is highlighted by the fact that 'the central monolith was broken during land clearing . . . and other stones on the site have been rearranged' (1992: 143).

Until 1993 Fiji had no Department of the Environment, resulting in the dispersal of responsibility for environmental management amongst fourteen ministries and related agencies. According the Director of the National Trust for Fiji (NTF), Birandra Singh, the total budget of the organisation for 1993 was a paltry F$80,000 (pers. comm.). According to the Corporate Plan (NTF 1992–3: 1), the Trust has wide-ranging responsibilities including:

1 the preservation of areas or items possessing national, historic, architectural or natural interest or beauty;
2 the protection of animal and plant life; and
3 the provision of access to and enjoyment by the public of such lands, buildings and chattels.

The NTF has a total of four rangers across Fiji, two of them full-time, one part-time and an 'honorary' ranger. The organisation prepared an evaluation of flora and fauna in the Mamanucas and recommended that the two islands of Monu and Monuriki be developed for day-tripping. The survey involved aerial and ground work and was intended primarily to promote the preservation of the crested and banded iguana. The Trust also sought Government assistance for the removal of feral goats, indicative of the scatter-gun approach to perceived problems.

Singh (pers.comm.) is not satisfied at the level of protection afforded in the Mamanucas and is critical of the resort operators, the dive companies and landowners for a 'lack of awareness'. He is optimistic about the potential for attracting tours to the Mamanucas focusing on environmental issues and cites the example of specialist groups which have come to watch migratory seabirds and sharks. He believes that the resorts could develop half-day tours to historic sites, particularly in times of poor weather. Good interpretation is needed. He believes that a properly resourced NTF should identify significant sites and be responsible for leasing them, with day-to-day management subcontracted to local landowners.

The National Environment Strategy questions the need for an enhanced role for the NTF (IUCN 1993: 51), commenting that 'the general failure of the Trust to significantly advance the conservation and preservation of Fiji's heritage, since it was instituted 21 years ago, is self-evident'. Despite the need for close integration of tourism into local communities, a strong national body is clearly needed which has the power to manage heritage assets effectively.

The objective of the National Environment Strategy was the achievement of sustainable economic development and resource use and the conservation of Fiji's natural and cultural heritage (IUCN 1993). It provides us with some baseline data on an area of activity (protection of the environment) regarded as increasingly important by visitors. It sought to bring about an effective environmental management capability, comprehensive heritage protection and meaningful private sector and general public involvement (1993: vii). Key issues that might be significant for tourism in the Mamanucas are the proposal that an official 'Register of Sites of National Significance' be developed (1993: x). It is also proposed that 'wildlife values be elevated within the public community through publication of appropriate material and awareness campaigns. This could be initiated by the official adoption of a national bird, fish or plant.' Such recommendations are indications of the potential for heritage and environment development in the Mamanucas. The fact that they are only now being canvassed indicates the lack of attention given to these significant issues to date. The report is also reticent on the need to involve the tourism industry in the development of environmental and heritage attractions, if such sites are to be made available and presented in a form that is attractive to tourists.

Heritage may manifest itself in a material sense as is the case with archaeological sites. It may also take a less tangible form through customs and behaviour. These dimensions of the Mamanucas' heritage assets are now considered in turn.

Archaeological sites

The deficiencies of environmental management in developing countries such as Fiji have already been mentioned. Because of resource limitations, one might expect the range of attractions to be less thoroughly documented than in equivalent developed countries such as Australia. Lack of documentation need not be an *absolute* deterrent to ecotourists, in that the very process of discovering about little documented features may attract the more intrepid travellers. Nevertheless, for the more typical ecotourist, a region must be known about to attract them rather than equivalent destinations. Some awareness and documentation of the relevant attractions is a likely prerequisite for this course of action.

Three major sources have investigated archaeological sites and monuments in the Mamanucas. These are:

1 the Fiji Museum's Sites and Monuments Record (SMR);
2 work undertaken by the archaeologist Dawson in 1987; and
3 sites recorded by the NLTB.

The Mamanucas are subject to the same deficiency as the rest of Fiji, namely the non-existence of a Register of the National Estate. The Fiji Museum's listing is the closest thing to a register currently available, but it lists just seven sites within the entire Mamanuca group. Listed sites are accorded no legal protection. According to Dawson (1987) only the island of Malolo has received 'detailed' archaeological investigation and even this study was confined to the coastal zone. The other deficiency is that much of the Mamanuca Group has never been surveyed by the Native Lands Commission. The map sheets and written records prepared by them have become an authoritative and informative point of reference for Fiji's archaeology and cultural tradition and could have provided valuable insights, had they been properly applied to the Mamanucas.

The Fiji Museum list records the islands of Tavua, Monu, Malolo and Qalito as possessing significant sites. Site types of natural and/or cultural significance include ring ditch fortifications, beach habitation sites, coastal features and valleys. Dawson's survey concentrated on coastal Malolo, though he discussed the possibility of Fijian archaeological sites occurring elsewhere in the Mamanucas group. For Malolo he identifies ten sites in total including five beach habitation sites, four coastal sites and one rock shelter. These sites showed evidence of early habitation such as shell middens and quarries. He believes that there are sites which would indicate early Fijian occupation on the islands of Tavua, Yanuya, Monu, Monuriki, Kadomo, Vanualevu, Navadra and possibly Malololailai. His judgement about early occupation is based on the presence of potable water and scale (the larger islands are more likely to have supported a resident population). Dawson believes that 'the habitable smaller islands extending northwest of the main Vitian islands sustained a sizeable prehistoric population extending back in time about 2,000 years' (Dawson 1987: 16). The NLTB survey identified an old village site on Vomo Island and *Vanua Tabu* (land whose access is limited by tradition) on Kadavulailai and Qalito ('Castaway') islands.

The optimum way to ensure that sites of regional significance are used for tourism purposes, is to adopt an integrated approach of site investigation, including the assessment of educational value and market attractiveness (for example, do they have appeal for domestic and/or international visitors?), creation of linkages between various individual sites (for example, trails), provision of markers, directional signs and interpretative materials and finally marketing. Important management functions include visitor impact analysis on site fabric and ongoing maintenance strategies such as the clearing of weeds, removal of litter and maintaining pathways to direct visitors' access to sites. To ensure that sites of regional importance retain their significance when used for tourism purposes, site investigation should conclude with a set of management recommendations ensuring that visitation to the site is sustainable, through the minimisation of access in sensitive areas. Some of these processes have been instigated for Malolo Island by Naitasi Resort. Working in conjunction with the Principal Tourism Officer (Ecotourism) in the Department of Tourism, the resort manager Max MacDonald and his wife Sandy are completing (at the time of writing) a brochure. The success or the failure of the venture will not become evident for some time, but the effort seems well directed. Awareness of the natural and human heritage of the Mamanucas is fragmented and piecemeal. If the social side of the Mamanucas is to be extended beyond the genuine smiling Fijian face depicted in promotional brochures, initiatives like that of the MacDonalds need both government and corporate support, if they are to extend beyond simple marketing exercises.

Myths, legends and cultural traditions

The potential of Fijian myth, legend and cultural traditions as a possible focus for tourism interest is confined largely to the northerly Mamanucas (Mamanuca-i-Ra) and to the Yasawas. There are currently no resorts in the Mamanuca-i-Ra area, though cultural traditions in the broader sense are evident in Fijian villages throughout the group. In a number of instances these traditions are already drawn upon by tourism with regular organised tours to various islands and villages.

The Mamanuca-i-Ra group features strongly in Fijian mythology and European accounts/interpretations of the early settlement of the Viti Levu mainland by migrant peoples arriving from the north west. Legend focuses around the arrival of the Fijian Vu, Lutunasobasoba and his followers, and the 'Naga Cult', a group of Melanesians adhering to religious beliefs and practices found elsewhere in Melanesia. They are of local and national cultural significance. In a regional sense, these legends have the potential to provide a context for various forms of tourism activity. Currently, the only available visitor publication covering the Mamanucas *The Mamanucas: Islands in the Sun* (Siers undated) focused primarily on the various resorts in the group and on the two 'pioneers' Dick Smith and Dan Costello. Whilst there is undoubtedly a place for such literature, the absence of any explanation of Fijian traditions and myths results in a very two-dimensional view of the area. It certainly fails to provide genuine distinction between the Mamanucas and comparable groups of island holiday destinations elsewhere.

The Narokorokoroko group (most of Mamanuca-i-Ra) is of great cultural significance to Fijians because it is 'the area where the Fijian ancestral Gods live' (NLTB 1980, Vol. 7). Cato (1950: 109) recounts that Narokorokoyawa island is the first place that Lutunasobasobo and his followers set foot in Fiji, prior to their going on to Vuda and the Fiji mainland. To the present day, anyone visiting the Narokorokoroko group is instructed to make a *sevusevu* (gift) to the ancestral gods at a cave on the island of Navanualevu.

There is little doubt that the heritage dimension of the Mamanucas has sufficient diversity to be potentially attractive to visitors. The desire of guests to be more fully involved in the culture rather than passive spectators at a *meke* was highlighted in a report by Stollznow Research (1990). From a regional point of view, the Mamanucas are well placed because of their powerful mythological (or otherwise) association with the founding fathers of Fiji. This offers the prospect for creating concepts to differentiate the region from other sun, sea and sand destinations.

Another aspect of Mamanuca heritage worthy of more thorough interpretation is the role of Malolo as the site of the first landing by a western military force on Fijian soil (Brown 1991: 50). In command of a US expeditionary circumnavigation of the globe between 1838 and 1842, Lieutenant Charles Wilkes launched a revenge attack on the village of Solevu following the murder of two crew members. Up to seventy-four casualties were recorded and the villages of Yaro and Solevu were destroyed by fire. Wilkes was subsequently court martialled for 'cruelty to natives'. Amongst the sites of potential interest to visitors are the anchorage of the invaders' boat *Porpoise*, the route taken by the landing party, the location of the sinking of two canoes and the place where the officers were struck down.

Another interaction between the US military and the Malolo villagers occurred during World War Two when, in 1942, Nadi Bay became an anchorage for the American fleet and Malolo island was used for training purposes (Brown 1991: 50). Other Pacific destinations such as Pearl Harbour in Hawaii have developed sites of military history into major tourism attractions. The Tourism Council of the South Pacific (TCSP) has identified 'Visits to World War Two Sites' as a major product development option for the Solomon Islands and for Tonga (Yacoumis 1992). The social and cultural links between Malolo and the US from 1845 to 1942 could be a worthwhile theme for visitor interpretation.

One dilemma in proposing to open up cultural heritage sites for tourism is that local residents are often unaware of the significance of particular sites. In other instances there may be a reluctance to recreate historic incidents, such as the burning down of Solevu and Yaro and subsequent 'abject surrender' (Brown 1991: 50) by the local chief, which are seen as humiliating. Sites of mythological significance are also considered to be very sensitive by local people. Organised tours have as a basic requirement that access to attractions will be provided as published, which constitutes a challenge for the management of some archaeological sites which may be sensitive to damage from over-visitation. Where access can only be determined on an arbitrary basis, product development will inevitably be curtailed. A field trip by delegates at a Tourism and Cultural Heritage Conference conducted by UNESCO in 1991 (and attended by the author) to various historic sites found one significant sacred site was too readily available, whereas landowners were hesitant and finally refused to allow entry to another site of lesser significance. Inconsistency of access could be a primary impediment to inclusion of sites within tour operator itineraries.

Another dilemma about opening up the local cultural sites to tourism is that it might overwhelm the (relatively small) local population. The existing 'sun, sea and sand' style of tourism is a form of enclave development, albeit a relatively subtle kind (that is, with low-key, traditional-looking accommodation which visitors may confuse as being typical of normal Fijian dwellings). One might argue that this provides the local inhabitants with an element of privacy, since relatively few tourists venture into the villages or explore sites of local significance. Others might argue that since much of the population of the Mamanucas is transient and has come to the area for employment, that attempts to present the 'true culture' of the Mamanucas are doomed to failure. The (largely undocumented) range of environmental heritage and cultural attractions would suggest otherwise.

This chapter has highlighted the physical appeal of the two island groups – an appeal which helped to propel the early development of tourism in both regions. The underlying (and to a large extent unexploited) socio-cultural and environmental appeal of the areas has been documented. In the next chapter, the actual form that tourism development has taken is examined.

3

TOURISM DEVELOPMENT –
ISLAND STYLE

Chapter 3 examines the historical development of tourism in the two regions, with particular reference to the role played by developers and entrepreneurs. Having evaluated the extent to which entrepreneurs have been willing to co-operate under the banner of collective regionalism, we proceed to a discussion of the political and institutional framework. To what extent has the institutional framework facilitated tourism development and to what extent have governments provided genuine support? Having an understanding of these attributes is essential as a backdrop to subsequent chapters on tourist activity in the various resorts. By concentrating on the islands themselves, chapters 2 and 3 have acted as a prelude to the two subsequent chapters on the tourism industry and on marketing. These later chapters focus more specifically on source markets and on their relationship with the respective destinations.

HISTORY: EXPANSION AND REDEVELOPMENT

The Whitsundays

The development of resort-based tourism in the Whitsundays predates that of the Mamanucas by several decades. According to Barr (1990: 7), island tenants in the 1920s supplemented their farming incomes by hosting local holiday tours and steamship passengers and 'by the end of the 1930s, approximately eight island resorts were catering for holiday travellers mainly within the winter trading season'. In other words, all Whitsunday island resorts were dependent on sea transport during their formative years. Relatively remote from the major population centres of Australia, isolation gave the Whitsundays a sense of exclusivity and even adventure. The ease of air access since the 1980s has ended the relative isolation, allowing inter-island sea transport to become an attraction in its own right. It is a challenge for those Whitsunday islands wishing to preserve a sense of remoteness and exclusivity to do so in the light of improved transport links.

As early as the 1930s, most island resort development was undertaken by

51

outside capital. Only two (Lindeman and Brampton) were developed solely by pre-tourist pastoral residents (members of the Nicholson and Bassutin families respectively). A key difference between the Mamanucas and the Whitsundays is that tourism developed initially in the latter as a means of supplementing pastoral incomes, though once the tourism potential of the group was revealed, outside capital was quickly forthcoming.

Barr (1990) has applied Butler's three 'stages of development' concept to the Whitsundays, using the stages: exploration (local residents establish experimental tourism enterprises, usually as a response to itinerant, self-generating visitation to the area); involvement (through the 1930s) and finally development where 'existing attractions became increasingly artificial and unrelated to the destinations's natural resources'. Though accurate government arrival statistics were not available for the Whitsundays prior to 1970, it appears that by the early 1960s the Whitsundays attracted similar visitor numbers to the whole of Fiji. According to Hundloe *et al.* (1989: 32), the islands attracted 5,000 visitors in 1947 and 28,000 in 1962. Through the 1960s and 1970s growth rates mirrored those of Fiji and according to Claringbould, Deakin and Foster (1984), grew from 28,000 in 1962, to 69,000 in 1969 and finally 182,000 in 1979. The comparable figures for Fiji were 18,255 in 1962, 85,163 in 1969 and 188,740 in 1979.

Local government played a major role in opening up the Whitsundays to tourism. In Fiji, developments were primarily at the instigation of entrepreneurs backed by the Fiji Government (previously the colonial administration). Barr (1990: 29) reports that 'in Proserpine, infrastructural reform was generated by the incorporation of tourism into the district's formal agenda for civic and commercial development'. This level of local government co-operation was not necessary in the Mamanucas, partly because of the easy accessibility from the Nadi International Airport. In the Whitsundays in the late 1940s the proposed construction of an airport at Proserpine aroused heated debate within council and within community organisations because certain residents such as W.B. Morgan rejected the high cost of the aerodrome and raised doubts over Proserpine's future as a major centre for tourism (quoted in Barr 1990: 30). An aerodrome was finally built in 1951 and was subsequently upgraded by the entrepreneur Reg Ansett who took control of the facility in 1957, marking the dominance of Ansett Airlines as a carrier into the region which was to continue for over forty years. Despite the gradual conversion of local decision-makers to the potential of tourism, the slowness of development did cause a period of stagnation for tourism in the Whitsunday mainland. During the 1950s problems with jetty facilities and with the state of Proserpine Airport caused the Queensland Government Travel Bureau (QGTB) to give preferential publicity to Mackay as the gateway to the Whitsundays. According to Barr (1990: 30), 'Mackay's dominance was confirmed in 1956 when the QGTB launched a major package holiday promotion of an eight-day tour through Mackay and the

Whitsundays'. Barr quoted the QGTB as writing that 'Mackay is the take-off point for the islands of the Barrier Reef on which there is a suitable standard of accommodation for overseas visitors' (Barr 1990: 30). Though an upgrading of facilities within Proserpine during the later 1950s enabled it to recover some of its position, the ebb and flow of Mackay and Proserpine as gateways clearly indicates that historical practice has not been over-concerned by strict geographical definitions of the Whitsundays. Practical and political issues seem to have been the major deciding factors. The tide is unlikely to shift back to Mackay Airport again. Proserpine Airport is now more commonly named Whitsunday Airport, indicating the facility's regional gateway status.

The Ansett strategy spanned both the mainland and the islands. Reg Ansett made major improvements to Hayman Island in the 1940s, but also developed Proserpine Airport and accommodation at Airlie Beach. The airline is currently owner of Whitsunday Connections which provides inter-island launch services and sightseeing trips to the Barrier Reef and to other sites of interest. Ansett executives have consistently argued that the risk taking and investment in Whitsunday tourism development by the airline in the post-war period justifies its monopoly of Hamilton Island Airport. Without mentioning the airline by name, current Ansett Managing Director Graeme MacMahon referred to 'other airlines which simply piggy-back risk taken by others' (*Travel Reporter* 1992b: 10).

The first visitor accommodation on Brampton Island was built in 1933 by Arthur Bassutin, the then leaseholder. It has been owned by Australian Airlines (now Qantas Airways) since 1985. According to Lamond's *Island Holiday* (1948), Lindeman was opened as a resort in 1929. By the early 1930s, owner Angus Nicolson had completed construction of cabin accommodation on the islands. From 1974 until the mid 1980s it was owned by P&O, indicating that large tourism corporations were active in Whitsunday Island resort management at an early date.

Daydream (or West Molle Island) is located 3 kilometres from Shute Harbour. In 1932 its lease was sold to NSW yachtsman Paddie Lee Murray who developed the island as 'Daydream Resort' in conjunction with Eric Catherwood. It was subsequently redeveloped by one of the pioneering companies of Australian tourism Cobb & Co, and was later purchased by Ansett (which also purchased Hayman). The resort has been twice demolished, once for redevelopment in 1952 by Bernie Elsey, a Gold Coast developer and then by Cyclone Ada in 1970. Its official name remained West Molle Island until 1989.

South Molle Island resort was developed by Ernie Bauer who became the leaseholder in 1937. The resort was until recently owned by Ansett Airlines, but in 1994 the lease was taken over by the hotel management company, the Jewel Group. Of the remaining islands, Hamilton Island claims to be the largest resort in the South Pacific. It has a large marina but is not dependent on boats because of the airport located on the island. Hayman was described as early as 1935 as a 'fishing resort' based on facilities developed by NSW teacher Monty Embury as a centre for scientific tours in the early 1930s. In 1985 it was closed, demolished and rebuilt as 'the premier international resort in Australia'.

What is now the Club Crocodile resort on Long Island was established as Happy Bay Resort and began operation in 1934. Another resort, precursor to the present day Palm Bay Hideaway, was developed in the mid-1930s by a Brisbane resident, Tim Croft, as Clear View Gardens. Southerners could arrive in the Whitsundays by one of two methods. Sailing the Whitsunday Passage was an essential component of the popular winter sea voyage from Adelaide, Melbourne, Sydney and Brisbane to Cairns. Southerners also arrived by rail. Visitors from the South were supplemented by a local clientele.

In the postwar period, Craik has asserted that until about 1970, the priority of tourism throughout Queensland was to attract motorists and that this constrained air-based travel to the Great Barrier Reef (1987a). Whilst she underestimates the significance of the early resort developments, she correctly identifies the constraining effect on growth caused by the remoteness of North Queensland.

Queensland is the dominant beach resort state in Australia. The Whitsundays constitute a middle-ranking destination in Queensland, with arrivals well below those achieved by Cairns, the Gold Coast and the Sunshine Coast. The Whitsunday Islands attract the highest average daily visitor expenditure of any Queensland region. Visitors from Victoria and NSW have made up an increased proportion of visitation and now account for over half of all visitor nights, though the balance differs greatly amongst the individual resorts. International visitor numbers have not grown as fast as interstate visitor numbers although their average daily expenditure remains higher. The Whitsunday Islands have increased their share of total Whitsunday arrivals relative to the mainland. Hamilton Island Airport has accounted for an increasing proportion of arrivals since its completion in 1983. The role of Proserpine Airport has declined over the same period. Inclusive tour packages are dominated by Ansett Australia and Qantas. Sunlover Holidays also features most of the resorts, as does Queensland Reservation Centre, a small and emerging operator.

The Mamanucas

In the early twentieth century, Fiji was the mid-way disembarkation point for ship-based passengers on the Australian–North America route. Fiji's first tourist hotel, the Grand Pacific in Suva, was completed in 1914. The Fiji Publicity Board (renamed the Fiji Visitors Bureau in 1953) was established in 1924. By the mid-1930s it was advertising in the *Bulletin* as well as some New Zealand publications, indicating that the Australian market was considered a potential source of business even at this early date. Scott (1970: 5) records that 'from 1946 to 1961, the Fiji Publicity Board was much occupied with demanding improvements at Nadi Airport and to national infrastructure and hotels'.

The first major report on Fiji's potential for tourism was the Checci Report (1961), which examined tourism as a means of strengthening the ·economies of the Pacific and Far East. By 1968, Fiji's annual visitation reached 66,467, exceeding the Report's prediction of 45,000.

The golden age of tourism from Australia to Fiji began in 1962 with the exempting of luxury goods such as cameras, telescopes and tape recorders from customs duty. This created the foundations of a 'duty free' industry. Hotel construction grew through the 1960s following the passage of the Hotels Aid Ordinance in 1964. Table 3.1 indicates the growth of tourism to Fiji between 1961 and 1993, outlining the relative market share provided by Australia.

Between 1968 and 1992, Australia accounted for one third or more of total visitors (it did drop to a little under 30 per cent in the early 1970s), making it the largest source of business. Australians have the longest average length of stay, enhancing their impact on the accommodation sector. Australia's share of total visitation in Fiji slipped to its lowest level since 1967 in 1993 (27 per cent in 1967, 27 per cent in 1993). This reduced share is not entirely negative since Fiji is broadening its spread of market sources. The second, third and fourth largest markets, namely Japan, the USA and New Zealand, accounted for 15 per cent, 14 per cent and 14 per cent respectively. This wide range of source markets is welcome because it reduces the susceptibility of the receiving country to changing economic conditions in one particular source country. Of more concern to Fiji is that Australian 'holiday departures' to Fiji as a percentage of all Australian overseas 'holiday departures' decreased from 10 per cent in 1982 to 6 per cent in 1992. If

Table 3.1 Tourism arrivals in Fiji 1961–95

Year	Total arrivals	Australia	Australian share (per cent)
1961–4	88,846	17,136	19.3
1965–8	207,377	55,478	26.8
1969–72	513,020	154,507	30.1
1973–6	697,342	237,648	34.1
1977–80	735,791	284,121	38.6
1981–4	818,418	362,797	44.3
1985–8	901,578	316,402	31.1
1989–92	1,086,629	374,317	34.4
1993–5	924,831	241,644	26.1

Sources: Scott 1970 and 1995

domestic air-inclusive destinations were taken into account, it is likely that the decline would appear even greater. Fiji's performance in the Australian market has clearly deteriorated.

Table 3.2 shows that a holiday is the principal purpose of a visit for over 80 per cent of travellers from Australia to Fiji. This confirms Fiji's role as a part of Australia's 'pleasure periphery' and suggests that Fiji's airline services will be determined almost entirely by the destination's continued ability to attract holiday travellers. If demand drops and capacities are reduced, there is a danger of a downward spiral, with consumers offered reduced flight and seat options. This has not occurred on any scale to date but is an example of the dangers of dependence. Fiji's failure to maintain its market share in the face of overall growth in Australian outbound holiday-making is a concern for the Fiji tourism authorities. Andrew Drysdale, Chief Executive of Air Pacific, recently asserted (Keith-Reid 1994a: 29) that airlines flying to Fiji had enough seats to land about 489,000 visitors a year (the current number of arrivals is about 300,000) . . . but, he went on to say, 'a lot of visitors prefer luxury hotels, not budget types. For several months of the year, the luxury places are full'. This comment is indicative of the importance of achieving a balance between the availability of airline seats and the supply of appropriate beds in the various accommodation houses.

Visiting from Australia to Fiji experienced a resurgence in the early 1980s when it accounted for 8–10 per cent of all Australian outbound holiday trips. This dropped to 5 per cent by 1993. The Mamanucas have traditionally enjoyed high occupancies, though these dropped to 73 per cent in 1993, a symptom of dependence on a declining Australian market (37 per cent of all Mamanuca visits). Room capacity in the Mamanucas is under 600 rooms with a further 600 located in the nearby Denarau Resort. Just under

Table 3.2 Holidays as a proportion of total outbound travel from Australia and of travel to Fiji from Australia

	1988	1989	1990	1991	1992	1993	1994
Total travel from Australia*	1.7	2.0	2.2	2.1	2.3	2.3	2.4
Total travel as holiday*	0.9 (53.9%)	1.1 (54.6%)	1.2 (54.2%)	1.1 (53.5%)	1.2 (51.4%)	1.2 (51.2%)	1.1 (48.6%)
Travel from Australia to Fiji	75,000	97,000	104,000	87,000	87,000	78,000	86,000
Travel from Australia to Fiji as holiday	58,600 (78.1%)	77,900 (80.3%)	82,300 (79.1%)	71,200 (81.8%)	73,200 (84.1%)	61,741 (79.6%)	66,546 (77.8%)

Sources: Australian Bureau of Statistics 1995 and FVB 1995
Note * Figures given in millions

a quarter of all international visitors spent at least one night in the Mama-nucas/Yasawas area. The area attracts many families, exhibiting a high ratio of persons per room. The largest properties are Plantation, Castaway and Mana islands. The area is also the predominant day-cruising area in Fiji. The key attractions of the region were cited as the people, followed by climate and scenery. 'One to two weeks' is the most common duration of trip cited by Australians and two out of three arrived on Air Pacific. Most properties are in the 100–200 room category, charging F$150–200 per night.

The NSW and Victorian markets are very significant for Fiji although, like other Australian sources, their numbers have dropped. Fiji (and pre-sumably the Mamanucas) has performed better in the market of families with older children and less strongly amongst teenagers and older age groups. In this respect, it compares unfavourably with Bali which appears to appeal to a wider range of age groups.

ENTREPRENEURS AND THEIR VISIONS

The Whitsundays

Hayman Island Resort and Hamilton Island are resort concepts which were driven by the 'vision' of particular individuals, respectively Sir Peter Abeles, the Managing Director of Ansett Transport Industries, and Keith Williams. Hayman Island was a well-established resort in Queensland having origin-ally been built in 1927. In 1947 it was purchased by Sir Reg Ansett, founder of Ansett Airlines and remained with the company under the name Royal Hayman Resort. The original purchase price was 100 pounds, or one pound for each goat on the island as agreed to by the previous lessees (Quorum 1986: 31). In the mid 1980s TNT–News Corp spent in excess of A$200 million to transform the island into an international five-star resort, with the objective of being one of the top five in the world. Ansett have not revealed the exact investment though estimates have ranged from A$200 million to as high as A$350 million (Rennie 1994). The resort was targeted at the top two to five per cent of the Australian population. Sir Peter Abeles' vision was to create a European-style concept in Australia. In contrast to Hayman's efforts to attract a predominantly international cli-entele, Hamilton set out to satisfy the Australian market. As Williams has stated, 'we wanted to have them holiday in their own country' (quoted in King and Hyde 1989b: 207). Williams, however, was not averse to borrow-ing overseas concepts, including 'Polynesian bures'.

In a sense, both Hayman and Hamilton are monuments to the develop-ment excesses of the 1980s. In 1988 it was reported that Hayman Island was losing in excess of A$20 million per year (Sandilands 1991b). The resort can never offer a return on the original investment to its owner and developer Ansett Transport Industries. If one relates the redevelopment cost of A$250 million to the 225 rooms, the development cost works out at A$1.1 million per room. Other estimates have ranged to as high as A$1.3 million (Sandilands 1991b: 94). This is far in excess of any other major hotel or resort development in Australia. Asked if Hayman really needed to charge A$1,000 a night per room to break even on the basis of a A$1 million investment per room, then Ansett Executive Director Alan Notley responded that 'this is typical of the shallowness of so-called analysts who work on the rule of thumb that for every $100,000 a room costs to build, you have to charge $100 a night. It doesn't work that way' (quoted in Sandilands 1991b: 103). Nevertheless the doubts remain. An analysis of Hayman estimated occupancy rates in 1991 of 45 to 48 per cent including non-revenue guests, a figure much higher than the travel industry generally accepts – and this is roughly a cash break-even situation where the takings pay for the daily running costs but not the servicing of the debt. If that is true, the real losses on the resort would be running at between A$30 million and A$40 million per annum (Sandilands 1991b: 103).

In 1991 the resort opened up to travel agencies for the first time, offering 50 per cent discount to *bona fide* travel agents (*Inside Tourism* 1991b). This may have been a sensible remedial policy, but it points to the failure of the initial belief that clients would pay rates substantially higher than at any equivalent destination and that the resort would avoid the need to pay commissions, overrides and substantial tour operator margins. Hayman now advertises through wholesalers and engages in deep discounting, just as its competitors have done. Such discounting cannot be put down to recessionary conditions – it is a major strategy failure, calling into question the research and policy capability of a major Australian tourism operator. Whilst Hayman's rates are the highest in the Whitsundays, the differential is relative and package prices of as low as A$799 inclusive of airfares and breakfasts have been offered by the resort in Melbourne.

An irony of the Hayman Island development is that its mastermind, Sir Peter Abeles, was forced from his role as Managing Director and Chief Executive of Thomas Nationwide Transport (TNT) and Ansett Transport Industries in 1992 and as Chairman of TNT in 1993. Amongst the criticisms levelled at Abeles were his high executive salary (several million dollars per annum) coinciding with low returns to his shareholders (through joint owners of Ansett Transport Industries, TNT and News Corp) (Ryan and Burge 1991). The 'visionary' development of Hayman which drained the resources of Ansett was a significant influence in the ultimate downfall of Abeles (Sandilands 1991b). At the same time, the positive outcome of Abeles' vision was that Ansett consolidated its dominant position in the Whitsunday Islands. The leasing of the A$16 million South Molle Island at the middle range of the market ensured that Ansett's Whitsunday focus was not too narrow. Though Hayman has been a financial debâcle for Ansett, it is evident that Ansett's focus on concentrating its resort leases in the

Whitsunday area has been a better strategic ploy than that of its rival Qantas (then Australian Airlines). The Qantas division Australian Resorts recorded profits for its five Queensland island properties in 1993 after six years of successive losses (Dowling 1993). When Qantas promotes its Queensland island resorts it is promoting the destination as a whole, including properties owned or leased by Ansett. In Ansett's case, its Whitsunday promotions can emphasise that the Whitsunday islands *are* Ansett.

Like Hayman Island, Hamilton Island is atypical of any other resorts in either the Whitsundays or the Mamanucas. This is firstly because of its large scale, with almost three times the number of rooms found in any of the other resorts. Secondly, it is the only resort island in Australia with its own jet airport and thirdly, because it was Australia's first resort designated as fully integrated when it was initially developed in 1984 (Millar 1992). This involved the private development of large-scale infrastructure and the provision of local services normally provided by local government. Controversy dogged Hamilton Island from the start, though this is not surprising in view of the vast scale of the endeavour. The subsequent resort owner and developer Keith Williams had acquired the 750-hectare island as a grazing lease and had introduced deer to the island to maintain the status of the lease (Williams 1988). However, the lease expressly forbade the development of tourism, raising questions as to how Williams managed to secure development rights for much of the island (Dickinson 1982: 39). Questions were also raised about the small amount for the lease payment (A$6,450 per annum according to Dickinson) and the secrecy with which Cabinet had changed the conditions of the lease (denied by then Lands Minister William Glasson). Further questions arose about the preparation of an Environmental Impact Statement (EIS) requested by Proserpine Shire Council and received in May 1981. The EIS was prepared on behalf of Hamilton Island Enterprises Pty Ltd by Ullman and Nolan, who were also the Proserpine Shire Council's engineering and town-planning consultants. According to Dickinson (1982: 40), 'Ullman and Nolan told the *National Times* that they knew nothing of a proposed airstrip for international aircraft'. The strident rhetoric of Williams' dealings with both government and environmental issues contrasts with the more diplomatic tone which prevails in the 1990s in the Whitsundays.

The Williams philosophy was very different from the types of values espoused in the current Draft Whitsundays Tourism Strategy. Whilst the Strategy espouses effective environmental protection, Williams bitterly opposed the inclusion of Hamilton Island in the Great Barrier Reef Marine Park (Marshman 1992). He also opposed the involvement of international hotel management in the operation of resorts. He claimed that 'not one resort in Australia has given investors a return on capital invested where it was run by an international chain' (*Inside Tourism* 1988). Williams' claim may have been symptomatic of the earlier practice of international chains to dictate the terms of the contractual relationship with property owners. The keenness of international chains to gain a foothold in the Whitsundays has reversed this relationship. According to property consultants Colliers Jardine, Holiday Inn's acquisition of the management rights to Hamilton Island was 'almost deferential' with a 'performance based contract emphasising the shift towards strict performance criteria and increased owners' involvement' (Colliers Jardine 1994: 3). Payment by Holiday Inn is now linked directly to the owners' overall financial return.

In practice, international chains now have a major presence in the Whitsundays (including Holiday Inn on Hamilton Island itself) and were viewed as a positive influence by all of the resort managers interviewed for the present research. Williams' insistence that Qantas should not be granted operating rights at Hamilton Island Airport put him at odds with the Whitsundays Visitors Bureau and with most of the other resort managers (apart from those directly associated with Ansett, of course). Finally, Williams withdrew from the WVB because of Council's opposition to the expansion of Hamilton Island resort and airport because it might damage Proserpine. He also argued that Hamilton was the hub of the Whitsundays and consequently did not need the WVB. In practice the relationship between Hamilton and the Bureau became increasingly cordial as Williams' influence waned.

Williams' 'vision' was primarily as a developer rather than as a manager. The incoming Holiday Inn management were highly critical of the management systems put in place prior to their arrival. Financial reporting systems were criticised as being inadequate or non-existent, recruitment as fragmented and marketing as over-dependent on Ansett Holidays and the Japanese market (Dowling 1994). On the other hand Williams' development of a rapid and efficient transport enabled him to 'reduce building costs on the island to approximately 30 per cent above comparable mainland prices. 100 per cent had been the previous norm' (Williams 1988: 198). This type of investment helped bring about a change in perceptions about how business could operate on islands. The construction of the airport was also visionary in that it was an astute investment (it cost approximately A$18 million). As Williams has stated, 'Civil engineers and other supposed experts in airport construction looked at my proposal . . . and virtually wrote me off as being something of a lunatic' (Williams 1988: 193). Though environmental concerns persist about the way in which the airport was developed, Williams' vision has undoubtedly enabled the Whitsunday islands to compete with otherwise more easily accessible mainland destinations such as Cairns.

If vision is judged on the basis of return on investment, most of the major Whitsunday developments have been unsuccessful, some spectacularly so. Daydream makes an interesting example. In December 1994 its owner, the Jennings Group launched a formal campaign to sell the resort. According to Smith (1994) the selling price will be approximately A$30 million. She reported that 'because foreign companies are prevented from buying 100 per cent of islands in the Whitsunday Region, Australian institutions will be the first approached to buy the . . . resort. The island was the subject of a $100 million rebuilding program in the late 1980s' (1994: 1). The fact that entrepreneurial vision does not often equate to profit is highlighted by the low profit margin offered by the resort. Smith has estimated that if the resort is sold for A$30 million and on the current profit estimate of A$2 million a year, the yield on the resort will be 6.5 per cent, well below the 10 per cent yield that institutional investors expect from their investments. The Daydream example which is fairly typical of Hamilton, Hayman and Lindeman is symptomatic of the fact that many resort developments in Australia are loss-makers for the first and second owners. Adequate returns on investment are often absent until the third purchaser has the opportunity to make a purchase at a heavily discounted price.

In view of the major losses sustained by developers in the 1980s, it is perhaps surprising that little resort development is currently underway or proposed for the islands. A number of projects are, however, underway on the mainland. They indicate that the trend on the islands is one of consolidation and not expansion.

The Mamanucas

Prior to the construction of the first Mamanuca resort, Castaway Island in 1966, the major activity within the island group was day cruising from Nadi and Lautoka. The scale of this activity had grown rapidly during the 1960s with visitor numbers to Fiji recording an increase from 14,722 to 44,561 and a substantial (though undetermined) proportion of these engaged in cruising activity. Dan Costello, developer of Beachcomber and Treasure Islands, began his venture in the 1960s as a day-cruising business from Lautoka. The current Managing Director of Turtle Airways and founder of South Sea Cruises, John Pettitt also began in the day cruising market, though in his case he remained in the transport sector and did not diversify into resorts (Keith-Reid 1994b). Both are Caucasians, as were all of the earlier tourism developers in the region.

Why did substantial resort developments occur on some islands and not on others? The commonest explanations are size, location, the term and nature of the appropriate lease and primary function (for example, resort, day trip destination, agricultural or other traditional activity). The islands may be categorised into three, namely:

1 relatively small scale resort development;
2 islands used as a tour destination (day-trip islands); and
3 those which are mainly in traditional use though some of the latter category have developed close relationships with individual resorts.

Development is concentrated in certain areas. Only two resorts, Mata-manoa and Tokoriki, are part of the Mamanuca-i-Cake group which lies relatively further from Nadi than two of the other groups and faces relatively poor accessibility. Varying levels of visitor contact are experienced by the islands which do not have resorts. There are important linkages between the inhabited islands and the (otherwise uninhabited) resort islands. Villagers resident on Yanuya island use Tokoriki island to do most of their planting – an example of integrated economic activity. It means that resort guests at Tokoroki may come into contact with Fijians not actively employed in the resort itself. There is also movement in the opposite direction, with Yandua island functioning partly as a day-trip destination for house guests from Mana, Castaway and Matamanoa islands. There are no resorts located in the northerly islands of the Mamanuca-i-Ra group, though one development has been mooted (unsuccessfully so far) for Kadomo Island. A notable feature of the Mamanuca-i-Ra group is its role as 'the area where the Fijian ancestral Gods live' (NLTB 1980). This aspect will be dealt with later in the chapter and points to the potential tourism significance of islands not currently used as resorts. It also points to the fact that the resort islands should not be regarded as being divorced from the adjoining environment.

Starting with Qalito (Castaway Island), the islands of the Malolo Group were the first to be developed for resorts. Though only one of the five islands has a permanent Fijian village (Malolo Island, which has two settlements), integration with the local community is greater than at first appears the case. Many villagers from Malolo Island can walk to the freehold island of Malolo Lailai (literally 'little Malolo') at low tide and work at one of the two resorts on that island, Musket Cove Resort or Plantation Resort. A second resort, Lako Mai, is currently being developed on Malolo Island. A sketch drawing of the resort has appeared in selected travel brochures for the 1994–5 season. The status of Malolo Lailai island as the only area of freehold land in the Mamanucas is significant in that the landowners are not obliged to adhere to the rules laid down by the NLTB in conducting relations with local Fijian residents. The land was an example of territory that 'had been *bona fide* bought by or given to Europeans and other foreigners before Cession' (Usher 1987: 89). This occurred because it had previously been purchased and operated as a copra plantation (hence the name 'Plantation Island').

Of the 'other islands' category, Tai (Beachcomber) and Navini are tiny as evidenced in table 3.3. Tavarua and Treasure are also relatively small, giving guests the impression that they are surrounded by other visitors. Another interesting aspect of the 'other islands' is the relatively wide range that are used for day-trip purposes. The purpose of such visits appears to be pure relaxation on some (generally the unoccupied ones) and cultural visits on others (usually islands with villages). Another dimension of islands across all four groupings is that there has been a variety of developments mooted for the area. Many of the smaller proposals supported by the NLTB have not gone ahead. The islands with the largest resort areas are most likely to grow further. Sheraton Vomo has plans for a second stage of development and major expansion is already under way at Mana Island. Table 3.3 lists the diversity of scale evident amongst the various resorts.

Whilst some resorts occupy and dominate tiny islands, Naitasi Resort is an example of a relatively small development on a much bigger island (Malolo which is 960 hectares in size). The native leases which affect all islands except Malolo Lailai are fairly stable with a number extending to ninety-nine years. While significant resort development has halted on most islands (Mana and Musket Cove resorts are exceptions with major expansions currently under way), there are still one or two projects either under construction or else proposed.

In the absence of any major physical expansion, it is clear that the arrival of a significant international hotel management company could only come about through acquisition. The recently completed first stage of the Sheraton Vomo Resort is an exception to this rule and the commitment of the ITT Sheraton Group to the 'Mamanucas' concept is yet to be evidenced.

Having identified the key role played by entrepreneurs such as Williams in the Whitsundays and Smith and Costello in the Mamanucas, we now focus on the institutional framework in Fiji and in Queensland. To what extent have the efforts of the entrepreneurs been complemented by the activities of the relevant institutions and organisations? The role of regional tourism structures is critical here and is worth a detailed comparative analysis.

Table 3.3 The area covered by the Mamanuca islands and resorts

Resort	Area (in hectares)
Plantation	9.3
Sheraton Vomo Resort	87
Mana Island	62.7
Tokoriki	11
Matamanoa	10
Lako Mai (under development)	6.1
Treasure Island	5.7
Naitasi Resort	4.1
Tavarua Island	4.1
Beachcomber Island	2.4
Navini Island	2

Source: NLTB 1980

THE INSTITUTIONAL FRAMEWORK

The Whitsundays

The Whitsunday Visitors Bureau (WVB) which has its headquarters at Airlie Beach, is a membership-based organisation which aims 'to facilitate and co-ordinate the collective interests and efforts of the Bureau's members' (WVB 1993: 2). Called the Whitsunday Tourism Association until 1992, the organisation recently added regional development to its existing destination marketing responsibilities. The Bureau is funded by QTTC, by the Whitsunday Shire and by members. It operates a Tourist Information Centre at Proserpine and distributes promotional material through other outlets and travel trade fairs.

The range of responsibilities allocated to the WVB is wide and its ability to manage activities apart from marketing is questionable. Despite the strong rhetoric expressed in recent business plans (WVB 1993) about the natural beauty of the Whitsundays area and the importance for consumers of product presentation and visitor service, the WVB is poorly staffed and resourced. The 1993 plan made no commitment to tangible environmental initiatives. Nevertheless, the Bureau appears to enjoy the support of the membership and produces regular strategic planning documents. Whitsunday residents and operators do have an opportunity to debate regional tourism issues through a formal process, albeit one determinedly pro-tourism. The type of regional tourism management practised in the WVB is positive relative to the much less developed structure prevailing in the Mamanucas.

The Whitsundays constitute one of the sixteen tourism regions in Queensland, co-ordinated by the QTTC's Regional Tourism Department and QTTC is represented on the WVB Board (QTTC 1993a). The sixteen Regional Tourism Associations (RTAs) received grant funds of A$1.710 million during 1992–3, supplemented by funds for participating in co-operative advertising activities. The responsibilities of the QTTC Regional Tourism Manager include RTAs and industry training, tourism and marketing research and Aboriginal and Torres Strait Islander tourism. All of these responsibilities are relevant to the research undertaken in the preparation of this book.

There is a relatively higher level of regional intervention by the QTTC than by its equivalent in other states such as Victoria. The recent appointment of a co-ordinator of Aboriginal and Torres Strait Islander tourism within the Regional Tourism Department is a potentially significant expansion of the role of the regional group. The QTTC corporate plan (QTTC 1993b) set out key regional strategies including:

1 improved regional business plans through the implementation of QTTC performance criteria;

2 develop co-operation between the various RTA's to create a 'zonal marketing concept';
3 to create co-operative marketing initiatives including 'agreed zonal initiatives' and national advertising campaigns; and
4 to improve industry quality and professionalism, product development and marketing through liaison with other relevant industry associations.

The formal nature of regional tourism in Queensland contrasts with the informal and voluntary structure prevailing in Fiji (for example, regional tourism in the Mamanucas is handled by a branch of the Fiji Hotels Association).

Since its establishment in 1979, the QTTC has been the most 'entrepreneurial' of the various Australian State and Territory Tourism commissions. It developed its ATLAS destination database and land content reservation system which enabled travel agents to obtain information about areas such as the Whitsundays and to make reservations. ATLAS was a fully-owned subsidiary until its sale to Telstra in 1994.

The QTTC also operates a wholesale tours department, 'Sunlover Holidays'. QTTC operates its own retail sales network, called Queensland Government Travel Centres. Sunlover Holidays have an impact upon the Whitsundays in two principal ways (S. Brewster pers.comm.). Firstly, Sunlover incorporates a number of smaller products or products in unusual combinations that might otherwise be neglected by the main (airline) operators. Secondly, it provides an opportunity for Qantas aligned travel agents to book clients through Hamilton Island airport flying Ansett, without actually booking an Ansett Holiday. The existence of Sunlover has lessened Queensland's dependence on the two domestic airlines to generate air-based holiday business (the Goss Government showed its desire to be independent of these two corporations by making an ill-fated A$10 million investment in Compass Airlines Mark 2). The wholesaling, retailing and reservations activities of the QTTC undoubtedly create marketing advantages for Queensland relative to Fiji.

The QTTC has been a highly politicised entity, notwithstanding the conduct of a number of 'independent' inquiries, notably the Kennedy Report in 1990. Ostensibly the various official evaluations of QTTC performance have been prompted by a desire to improve efficiency and by the desire of politicians to portray the QTTC as free of government and bureaucratic interference. The actual effect has been to subject functions such as research and regional tourism to prevailing (and often political) preoccupations.

A legacy of the Kennedy Report was to increase the emphasis on visitor growth targets and accountability. A corporate plan, *Vision 2000* set out ambitious targets for the millennium. Between 1990 and 2000, visiting from NSW was targeted to increase by 117 per cent and from Victoria by 142 per cent. If such growth were to happen, Queensland's share of the domestic market would rise from 21 to 34 per cent. International visiting was set to increase faster still and to exceed interstate visits by the year 2000. Such figures provide politicians with welcome news and they are convenient because most politicians are no longer in office by the time they are shown to be grossly over-optimistic.

In seeking to balance the expectations of the various Queensland tourism regions, the QTTC funds a brochure covering each region using a uniform presentation. The QTTC can determine whether to increase or to decrease its financial support for a regional tourism organisation through its efficiency measurement mechanism. The third method affecting the operation of regional tourism is co-operative promotion and advertising where individual operators work through the regional tourism associations to secure QTTC funding. These mechanisms are, on paper at least, more sophisticated than anything in place in Fiji.

In practice, more developed tourism regions attract greater QTTC funds because large operators are able to raise promotional dollars which in turn can be used to attract dollar for dollar funding. Those Whitsunday islands managed by international hotel companies are major beneficiaries, since they have enough funds of their own to undertake television advertising (such advertisements usually begin with an overall Whitsundays promotion followed by reference to a particular property). Smaller operators are often confined to less costly promotional media such as radio and print. The fact that the Whitsundays span a highly developed area (the islands with their large promotional budgets) and an area perceived as lacking in tourism product (the mainland) allows it to be a major beneficiary of QTTC support. In contrast to regions such as the Sunshine Coast which are regarded as suffering from a lack of local co-operation (*Travel Reporter* 1993a: 2), the Whitsundays with their improving display of co-operative activity are seen as a sort of advertisement for the QTTC's attempt to strengthen the tourism regions. To this extent they should not be regarded as typical of what occurs in Australia's island resort regions.

A Queensland-based cultural policy specialist, Dr Jennifer Craik has criticised the Queensland Government for operating 'in a "policy vacuum"' and for 'delegating too many responsibilities to the QTTC which was not equipped to deal with policy related issues' (*Travel Reporter* 1993b: 4). The Queensland Government subsequently signalled a more proactive role for Government departments in tourism strategy development. It created a Tourism Policy Unit within the Queensland Government's Policy and Legislation Unit and established close links with the Department of Tourism, Sport and Racing (DTSR). This allocation of responsibilities has released the QTTC from the need to 'develop new Zonal Marketing Plans' for each of the regions (QTTC 1993b) though continued emphasis was envisaged for monitoring the quality of the regional Tourism Business

Plans. Other changes included the enhancement of research which had been a low priority in the Vision 2000 document. Regional Tourism was no longer regarded as simply a conduit for data from the QVS and could 'undertake . . . relevant domestic marketing research' and provide and interpret 'tourism research from internal and external sources' (QTTC 1993b). The close relationship between regional tourism and tourism research offers resort operators in the Whitsundays the opportunity to obtain appropriate marketing information. This could provide operators with a better understanding of their current and potential customers.

The WVB and the individual resort island managers (D. Hutchen, S. Cogar and K. Collins pers.comms.) all acknowledged that commitment to regional co-operation within the region began only in 1991–2. Most resort general managers perceived benefits in the regional approach for their own property as well as expressing more altruistic ambitions for the region. Vincent (pers.comm.) described it in the following terms:

> Destination marketing will benefit Daydream more than any other resort. First they choose the Whitsundays, then they can choose between Hamilton – high rise, Hayman – too expensive, South Molle – too old, Club Med – don't know what it is, don't know if I really want to go to Club Med and it is more expensive, and Laguna Quays – golf resort. We fit best.

Convincing managers that self-interest and collective interest can co-exist is probably the best recipe for effective regional marketing. Given the fairly recent history of genuinely co-operative activity, one must be cautious in making assumptions about an ongoing commitment to co-operative activity, however strong the apparent logic for such an approach.

The politics of Whitsundays tourism is complicated by the existence of two distinct areas within the region, namely the mainland with its predominantly smaller operations and (in the case of Airlie Beach) backpacker-style facilities and the islands with their image of exclusivity and the prevalence of international hotel brands operating the resorts. Almost all of the residential population is on the mainland and makes up the electorates of the councils of Pioneer and Whitsundays which are in turn members of the WVB. As is the case elsewhere in Queensland (B. Wallace pers.comm.), councils seek to influence regional tourism development arguing that their case has the backing of thousands of electors. The presence of a QTTC representative on the Board of the WVB imposes some uniformity on regional tourism across the State.

The island resorts have the largest marketing budgets in the region whilst the strength of the mainland is its electoral and administrative base. Compromise does not necessarily lead to beneficial planning outcomes, but in the case of the Whitsundays the different interests are harnessed through the Whitsunday Tourism Strategy (still in draft form in mid-1996). Management of the strategy development process by the Office of the Co-ordinator General (located within the Premier's Office) ensures that the Strategy has the highest imprimatur of the State Government. According to the Tourism Policy Unit within the DTSR (S. Wall pers.comm.), co-ordination through the Premier's Office gives the strategy greater impetus amongst government departments at the implementation stage. Involvement of industry (through the WVB) and local government (through the Pioneer and Whitsunday Shire Councils) as the main participants in the process along with the State Government encourages a co-operative approach to the resolution of competing sectoral interests. In view of the recent change of government in Queensland, the future of these arrangements is currently uncertain.

In practice the outcomes of the strategy do not appear to polarise between mainland and island interests. The image development for the Whitsundays proposed in the draft Strategy consists of three elements, namely 'an island/Reef/marine theme for the offshore area; a coastal theme emphasising the land/sea interface for the Whitsundays coast and a rural theme for the Proserpine hinterland' (Office of the Co-ordinator General 1994: 27). Figure 3.1 provides a promotional example of that strategy. The absence of resort development opportunities in the islands due to environmental constraints, may shift the regional balance towards the coast and reduce any residual bi-polarisation.

A number of opportunities arise from the incorporation of the mainland and the islands within a regional approach. Most specifically, it can allow the islands to develop a stronger sense of place. Currently the islands emphasise their proximity to the Great Barrier Reef and the classic imagery of the 'tropical island paradise'. Greater emphasis on regional concepts, however, can provide the destination with a stronger sense of integrity and authenticity. Example of the types of integration recommended in the draft Strategy are the encouragement of 'links between resorts and mainland to improve the use of complementary services, and shopping opportunities to include local products for the emerging local arts and crafts industry', to 'facilitate multi-resort visits' and to 'encourage linkages between resorts and national parks regarding visitor use and support' (Office of the Co-ordinator General 1994: 32).

Another advantage of the integrated strategic approach to tourism development is the integration of tourism marketing with the realities of physical and local government planning. This is particularly important in the Whitsundays where two separate local government strategic planning schemes apply, as well as the large number of national and marine parks included within the Great Barrier Reef World Heritage Area.

The predominant resort manager perception of relations between the islands and the mainland is dominated by the view that the islands have recently 'got their act together' and have thus been able to wield the

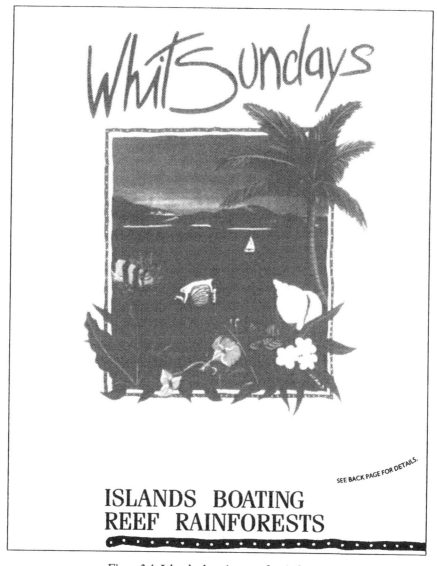

Figure 3.1 Islands, boating, reef, rainforests
Juxtaposed images may prompt consumers to form mental associations between the mainland and the islands, though references to the Whitsunday 'rainforest' may invite unfavourable comparison with Cairns.

influence commensurate with their marketing budgets. The lead role of the island resorts is acknowledged by the Chairman of the WVB (D. Hutchen pers.comm.) who commented that the role of Airlie Beach is primarily as a service town.

Bill Vincent of Daydream Island Travelodge Resort (pers.comm.) regards the lead role played by the islands as a function of available marketing budgets. 'The islands have more money to spend. You get criticism that everything is the islands and that the mainland gets forgotten. That will not change unless you get another major player coming into Airlie Beach.' The exception to this is the golf resort, Laguna Quays located on the coast south of Proserpine, whose management was recently assumed by Southern Pacific Hotel Corporation (SPHC), the operator of Daydream Island. Speaking prior to the SPHC assuming management of the property, Vincent stated that 'whenever we do anything as an island group, we always include Laguna Quays'. He commented that the Whitsundays area is unusual in that 'six or seven main players make up 90 per cent of the hotel rooms and market share. In co-operative marketing the locals see the resort names up there more often, but only because we spend more money'. The involvement of SPHC on both the mainland and the islands is likely to lead to further co-operative marketing activity.

Vincent's view of the significance of the Great Barrier Reef as an attraction is at odds with some of the other managers. In contrast to Klein (pers.comm.) who stated that 'for the domestic market the Great Barrier Reef is a natural drawcard for us because we are the island closest to the Reef', Vincent's view was that it is

> not significant as a lure to the domestic market – only to the international. People down south associate the Great Barrier Reef with Cairns and not with the Whitsundays. Certainly we would like them to associate it with the Whitsundays since we have just as close an association as Cairns does and we have the advantage of many offshore islands.

In response to the criticism that the entry price for the co-operative marketing activity is pitched too high for smaller operators, Stephen Gregg (pers.comm.) commented that the entry point for television-based advertising was obviously high, but that smaller operators still have access to the scheme using lower cost options such as print-based marketing. From a regional point of view, it is important that the images projected by individual operators are linked closely and consistently to the region as a whole. The co-operative marketing scheme enhances the likelihood that this will be the case. The absence of such a scheme in the Mamanucas may have held back the emergence of a more co-operative approach by the resort operators.

Managers at the smaller Whitsunday properties of Hook and Palm Bay said that they had little contact with the WVB. The relative lack of success in drawing the two smallest resorts into the co-operative approach should be acknowledged by the WVB. Though offering only small-scale capacity, both resorts offer an extra dimension to the Whitsundays product range.

According to Hutchen (pers.comm.), another property deliberately opting out from regional marketing is Club Med. The resort does, however, appear to maintain cordial relations with the other resorts, attend the monthly meeting of island resort managers and co-operate on environmental and related matters.

At Hayman, Klein was keen to bring about a strengthening of the mainland on the basis that the resulting benefits would impact upon both the region and the islands.

> We need more product on the mainland and need to change the image of the mainland from backpackers. We need to change the mainland from a pass through place to somewhere that you can actually stop. We need to make the mainland beaches cleaner, to spend more on improving the landscape and to make the mainland more of a destination in its own right.
>
> (Klein pers.comm.)

The need for greater product diversification on the mainland was also asserted by Sunlover Holidays (S. Brewster pers.comm.). Despite the extensive area covered by the Conway National Park, he regarded the shortage of organised product as a 'major impediment' to ecotourism in the area.

Cogar (pers.comm.) regarded the high level of representation of both island and mainland properties on the WVB as a positive indication of co-operation. He mentioned the representation of Laguna Quays and Club Crocodile (both mainland resorts), of significant boat operators and of the Chairman of the Whitsunday Council on the Board of WVB as evidence.

Whilst mainland interests claim the exercise of excessive influence by the islands, there is a perception that mainland interests can disadvantage the islands. Relations between local government and the resort islands appear to be closer than in the past. Cogar indicated that Council meetings are increasingly hosted by islands properties and that anti-development sentiment within Council is declining.

Brampton is the one resort which has not really benefited from the process, as recognised by Mahony (pers.comm.) who supported the regional concept saying that 'the region should market itself as a destination and everyone should support that. Once they are here, then the property with the best advertising and infrastructure will get the largest percentage of guests, but I think it is important that they all work together cohesively'. Clearly the success of the Whitsundays advertising has some concerns for Brampton since Mahony acknowledged the Whitsunday properties as constituting the main competition: '90 per cent of the Whitsunday islands we would see as our competitors'. Though the high level of cohesion is clearly a recent development, the Whitsundays demonstrate an impressive awareness of collective interest. In fact many of the adjacent areas wish to be associated with the Whitsundays name.

The Mamanucas

The tourism planning framework in Fiji is developed and implemented largely at national level. National tourism plans provide a perspective on where the Mamanucas are placed relative to national policy and the policies of NLTB offer insights into the relationships between native landowners and tourism developments.

The *Tourism Development Programme for Fiji* (UNDP 1973) was one of the first comprehensive tourism development plans focusing on a developing country that was funded by the United Nations Development Programme (UNDP) and World Bank. A Tourism Masterplan was also completed in 1989 and a subsequent plan is under way in 1996. The NLTB has also undertaken some studies and has prepared five-yearly tourism policies.

The 1973 Tourism Plan provided a sketch of how tourism would develop in the Mamanucas. It evaluated the key strengths of the group in terms of natural amenities, notably 'good beaches, clear water, interesting reefs . . . outstanding views', its 'small scale islands' and easy accessibility in view of its 'proximity to Nadi and Lautoka'. It also commented on the 'opportunity for a wide range of water activities' (1973: 102). Because of the small scale of the Mamanuca Islands it urged appropriate 'density and design'. It expressed a preference for development on those islands located relatively closer to Nadi and to Lautoka on the basis that a proposed 'high-speed boat service such as the hydrofoil type' (1973: 103) would be able to reach these islands easily. In fact, development has extended to some more remote islands such as Tokoriki and Matamanoa. As predicted, such islands experience accessibility problems and are not serviced by the hydrofoil.

The 1973 Plan anticipated the islands best suited to development including Qalito ('Castaway Island') and Tai ('Beachcomber') and the development potential at Mana, Levuka ('Treasure Island') and Vomo. Further development was encouraged for Malolo Lailai because of good accessibility available through the airport. The Plan featured an attractions map, urging the development of seven of the Mamanuca islands as nature reserves – Malamala, Monuriki, Monu, Kadomo, Vanualevu, Navandra and Eori. This proposal, though never properly implemented, indicates a realisation as early as the 1970s that the development of parks and reserves could be a valuable way of protecting the attractions base of the region. It also suggests that the 1973 Plan did not confine its vision of the Mamanucas to a purely 'sun, sea and sand' destination.

The Plan urged the further development of low-density *bure* style accommodation, compatible with the small scale of the islands. It proposed that hotels and resorts should be set back 100 feet (30 metres) from the high-water mark and that public access be provided to beaches adjacent to tourism development (this recommendation was subsequently incorporated into the General Provisions of the *Town Planning Act* (1980) (IUCN 1992:

82)). While the 1973 Plan probably contributed to the sun, sea and sand focus of the Mamanucas, it did propose a number of initiatives which, if adopted, could have resulted in a broadening of the concept. Tourism, it stated, should be used as 'a mechanism to establish national parks and nature reserves (terrestrial and marine), scenic areas and other points of natural interest' (IUCN 1992: 82). Unfortunately this proposal was never properly implemented in the Mamanucas (nor anywhere else in Fiji).

The Managing Director of UTC Tours, one of Fiji's major inbound tour operators and the largest for the European market, was dismissive of the 1989 Tourism Masterplan, criticising the inaccuracy of the statistics used and the failure to engage in extensive industry consultation (P. Erbsleben pers.comm.). The former Director of the Department of Tourism (M. Gucake pers.comm.) confirmed that 'the industry didn't like the 1989 Masterplan'. His own view was that 'presentation' was the problem. The proprietor of Musket Cove was also scathing of the report (Dick Smith pers.comm.), claiming that it over-emphasised what it regarded as potentially negative social and environmental impacts in the Mamanucas. According to the Plan, parts of the Mamanucas had surpassed an optimum ratio of guests to hosts (one to two) and 'it is recommended that appropriate surveys and evaluation be undertaken in this important tourism region to . . . serve as a case study for possible replication elsewhere in Fiji over time'. A 'prohibition on new resort development in some areas' might become desirable. It also urged research into the 'fragile physical environment' of the Mamanucas 'to ensure sustainable usage' (Coopers and Lybrand 1989: 139). Some of the recommendations may have been heavy-handed and unpopular with industry, but the perspective of the 1989 Masterplan on capacity limits and the pitfalls of mass tourism should be heeded in the 1996 Plan being undertaken by the Tourism Council of the South Pacific. It does appear that the 1989 Plan regarded resorts and the adjoining community and environment as being in frequent conflict. Arguably there was insufficient emphasis on the building of strong linkages between resorts and their surroundings.

The Fiji practice of landowner involvement in tourism development and activity has been regarded as a useful model by some other South Pacific countries. Between 1940 and 1946 the Commissioner of Lands, Ratu Sir Lala Sukuna investigated, defined and recorded the boundaries of Fijian landholdings and the NLTB assumed the responsibility for resolving any disputes and interpreting the boundaries, thereby minimising protracted wrangling which can impede tourism developments in certain other Pacific countries (Usher 1987: 71). Given the complexity of administrative and social structures within the Mamanucas, there can be no doubt that the NLTB has provided a useful mediating role, though its influence on day-to-day relations between communities and resorts is very limited.

As in the other Pacific countries, Fijian social, cultural and environmental issues are closely intertwined. The word *vanua* is typically translated as 'land' in English, but its true meaning shows its combination of culture and environment. 'The Fijian term, vanua, has physical, social and cultural dimensions which are inter-related. It does not mean only the land area and the vegetation and animal life . . . it also includes the social and cultural system' (Ravuvu 1987). The NLTB was established under the *Native Land Trust Act 1940* and administers native title land on behalf of the landowners. Typically a lessee of native land pays a proportion of gross turnover to the NLTB which in turn distributes this money to landowners, after deducting an administration fee. In 1990 monies distributed to native landowners for hotel and resort developments amounted to F$1,650,000, or 22 per cent of all funds distributed. This amount was third only to agricultural and residential leases.

Eighty-seven percent of the landmass of Fiji is classed as Native Land. Currently 43 per cent of the country's total current guest rooms are located on native land and it is anticipated that this percentage will increase. According to the NLTB (1990), 78 per cent of 'planned' rooms are on native land. As the Report explains, 'as the availability of suitable freehold land declines, the importance of Fijian-owned land as a source of supply for tourism development sites increases'. Native land accounts for 88 per cent of the accommodation stock in the Mamanucas/Yasawas and is higher if the Yasawas are eliminated from the calculation. Historically the NLTB has played a vital role in the Mamanucas. Following the identification of the Mamanuca-Nadi-Coral Coast region as the key development priority in the 1973 Tourism Plan, the NLTB proceeded to negotiate the leases for all of the resorts in the Mamanucas, accepting the process of expansion.

The role of the Board is set to change. Its exclusive emphasis on the Mamanucas-Nadi-Coral Coast region has been superseded by a commitment to allow more development in 'Type B Visitor Accommodation Regions' and in regions previously designated 'Visitor Interests Area'. These are areas which have not previously attracted substantial development. The NLTB view is that 'it is not considered desirable to allow over-concentration' (United Nations Development Programme 1973: 2). The report states further that 'there is wide feeling and recognition that further development in these regions [Type A Regions including the Mamanucas] could lead to a down-grading of the attractiveness of the tourism resource and unacceptable social costs and problems'. Consequently the Board is committed to a more decentralised approach including the 'identification of viable and available tourism development sites in other regions as an essential ingredient in any efforts to re-direct the flow of visitors within Fiji in the future'. The danger of this attitude, from the Mamanucas' perspective, is that NLTB may over-emphasise the expansion of capacity, possibly at the expense of a commitment to improving linkages between Mamanucas resorts and the adjoining environment and people. The changing emphasis of the NLTB away from 'area one' locations such as the Mamanucas to more remote parts of the country is of particular concern in light of the relatively small amount spent by the FVB for marketing Fiji in Australia, relative to marketing by other national tourism organisations.

The NLTB is heavily committed to a number of landowner-based pilot projects across Fiji but none of these is currently in the Mamanucas. It is also committed to a 'Regional Planning Approach to Tourism' involving the development of tourism resource information systems. For these, it does not intend to focus on the Mamanucas, instead emphasising the Yasawas, Taveuni, Kadavu, Lomaiviti and Lau. In the absence of an active interest by the NLTB in Mamanucas product development, responsibility will be shouldered by the resorts. The Board already acknowledges 'visitor numbers drifting away from the Mamanucas to the Yasawas and eastern destinations such as Taveuni and Kadavu' (United Nations Development Programme 1973: 4). It may be, however, that action is needed by the NLTB to assist the Mamanucas develop more of these environmental and cultural attractions. The need for more such attractions is acknowledged in the Report which states that 'every major report and consultancy to government on tourism has stressed the importance of identifying and promoting Fijian cultural and environmental attractions' (United Nations Development Programme 1973: 4).

From the 1960s to the present, the Mamanucas have been one of the two major tourism destinations within Fiji, though current government policy and market trends in Australia are tending to work against the group. The absence of any public funding for a regional tourism organisation appears to have hindered the ability and/or willingness of the resort managers and other representatives to act collectively. Some have supported the Mamanucas branch of the Fiji Hotels Association but the initiatives have not been consistent, testimony to inadequate resources.

There is a feeling in the Mamanucas that, after enjoying the highest occupancy rates in Fiji for many years, the group is now starting to lose ground with occupancies down by 12 per cent in 1993 over the previous year (Dick Smith, quoted in *Traveltrade* 1994). A number of trends have tended to reinforce this view. Government has been pushing so-called 'secondary tourism' activity, much of it away from the major tourism centres. There is an attitude that the very small operators which typify secondary tourism ventures are not really viable in the Mamanucas because the area is so dominated by resorts and by resort managers. The various pilot projects for landowner-based tourism projects sponsored by the Fiji Government and by the Tourism Council of the South Pacific are all located far from the Mamanucas. The Mamanucas are regarded as somewhat peripheral to the 'new' type of sustainable tourism. Though the benefits to the remote parts of Fiji have also been questioned (M. MacDonald pers.comm.), promotional campaigns such as 'Discover the Fiji You Don't Know' may have attracted the attention of potential visitors away from the Mamanucas, since the wording of the advertisement 'that you don't know' would appear to be referring to lesser-known parts of Fiji and not to the Mamanucas. Finally, the expectation by government that little growth can be anticipated out of the Australian market threatens the Mamanucas, unless the region can either attract an increased share of the existing Australian market for Fiji, or else can rapidly attract business from the faster growing long-haul markets.

The Mamanucas have been a classic sun, sea and sand tourism destination offering friendly Fijian service and some elementary exposure to the culture through activities such as mekes and kava drinking. It has offered plenty of water activities for visitors, but has not been seen as sufficiently diverse to cater for specific target markets. To date the potential of groups with a particular interest in the environment or in culture has been ignored. It has been shown that the Mamanucas have real potential to develop further their natural and cultural heritage assets including archaeological sites, village visits and folkloric interpretation as well as natural features such as reef trips and flora and fauna trails. Despite the identification of a number of potentially significant tourism assets in key environment reports, the fact that the Mamanucas are a low priority for the location of formally designated protected areas is of concern. Though environmental management issues are becoming better documented, the lack of resources to implement policy also acts as a barrier. A stronger institutional framework will be needed if the crucial synergies between resort-based tourism, environmental and cultural heritage are to be properly exploited.

In contrast, the balance between resort tourism and the environment in the Whitsundays appears to be in safer hands. The existence of a fairly strong and established regional tourism association and management plans for both tourism and the environment, has reduced the dependence of the region on quirky (albeit highly creative) entrepreneurs. There are, however, large clouds on the Whitsundays' horizon. First is the continuing neglect of socio-cultural heritage to create a distinctly regional concept. Second is the failure to convert important management plans from draft to final stage. The apparent gains that the Whitsunday islands have made and which are identified throughout this book will remain tenuous until the key tourism and environment plans are being genuinely implemented.

Part II
THE INDUSTRY

4

GETTING BUSINESS FOR THE PLEASURE PERIPHERY

This chapter evaluates the activities of the key tourism industry players influencing travel by Australians to the Whitsundays and the Mamanucas and the relative competitiveness of the two regions. Similarities and contrasts are brought out between outbound (that is, international) air-inclusive holiday travel from Australia and domestic air-inclusive travel. The statistics on flows and visitor profiles presented in the two previous chapters are compared and the current and potential selection of competitive strategies by industry players and by the destinations are assessed. Special attention is paid to industry structure, to the regulatory framework in which it functions and to the roles of the various players. Whilst the Australian market has a number of distinct characteristics, the chapter provides a useful point of comparison for 'pleasure peripheries' in other parts of the world. It raises key issues about the relationship between short-haul destinations and source markets, and about how such relationships are being influenced by changing industry structure.

Though Ogilvie's *The Tourist Movement* (1933) and subsequent studies have incorporated tourism flows and statistics and descriptions of 'industry' components such as travel agents, tour operators and the accommodation sector, analysis of comparative structures and interfirm comparisons have been lacking. The latter type of study has been confined largely to the management consultancy domain and rarely available to independent researchers, though the recent growth of specialist tourism marketing journals may alter this situation. The most thorough examination of the definition and scale of the tourism industry has been undertaken by Leiper (1979). Leiper (1990b) has argued that much of tourism is only partially industrialised and that the size of the tourism industry, as traditionally defined, is exaggerated. The distinction that he draws between generating, transit and destination regions within an overall tourism 'system', is a useful framework for comparative purposes. In the present chapter, mention is made of these distinctions. The different roles of transit and destination regions are very important for the viability of island resort groups.

A number of reports published by the London-based Economist Intelligence Unit are relevant to the present research, including a study of outbound travel from Australia and New Zealand (King 1992a), an overview of tour operators and air inclusive tours in Australia (King 1991), an evaluation of the Fiji tourism industry's response to the 1987 military coups (Scott 1988) and finally an overview of air transport in the South Pacific (Bywater 1990a). These reports provide an international context for particular industry sectors and market segments but do not engage in detailed comparisons of similar and/or equivalent areas. Possibly the closest to a detailed study on multiple destinations from the point of view of a single source market is McDonnell's *Leisure Travel to Fiji and Indonesia from Australia 1982 to 1992: Some Factors Underlying Changes in Market Share* (1994). McDonnell's study is useful for its attempt to examine comparative performance.

Other studies have examined the role and function of the package tour from various perspectives including spatial aspects (Pearce 1987b), package components (Sheldon 1986; Quiroga 1990), tour operator imagery (Reimer 1990), choice modelling (Sheldon and Mak 1987), segmentation (Thomson and Pearce 1980); and price bundling (Hooper 1994). Guitart's (1982) examination of package tours in the Mediterranean region provides a useful regional perspective. Most academic writing has analysed consumers in Europe and North America. Unlike the situation in Australia, it is still the case in Europe that most packages are rigidly structured around charter flights, often using aircraft owned and operated by the tour operator. Though changing, there is still a big contrast with air-inclusive holidays from Australia where looser structures favour consumer freedom of choice. Studies such as those by Access Research (1990) and by King (1991) are unusual in focusing on Australia.

The theory of centre–periphery relations is a feature of the development studies literature, observing the dependence of destinations upon decisions taken in metropolitan tourism-generating countries, typically within the developed countries. Britton (1980) and Samy (1980) have argued that Fiji is at the mercy of a number of powerful travel organisations based in developed countries such as Australia. Britton regarded tourism as an agent of neocolonialism in the former British colony and observed that tour operators can exert undue influence over sovereign authorities through their ability to divert tourism to destinations in alternative countries at short notice. In a crude sense Queensland destinations are also dependent on corporations based outside the state. Most of the travel agents and tour operators which promote Queensland (like those offering Fiji) are based in Sydney and to a lesser extent in Melbourne (King 1992a). Constitutionally though, Queensland is the equal of other states in Australia and the term 'neocolonialism' is not applicable. One might, however, make the case that Queensland is peripheral to the major metropoles and that the term 'centre–periphery relations' can be applied to tourism. The state certainly fits Turner and Ash's (1975) 'pleasure periphery' concept used in the title of the present chapter. Such a relationship can lead to a culture of dependency.

Studies of comparative advantage have also been used to assess the effectiveness of tourism destinations (Grey 1970). Porter's (1980) work on competitive strategy and competitive advantage (1990) has distinguished between the activities of individual firms in pursuing business opportunities in global industries and the criteria of national advantage which often determine such success (Glaister 1991).

Attempts to apply Porter's theories to tourism have met with limited success. Porter's (1980) view that overall cost leadership, differentiation and focus are the essential elements of competitive strategy has been rejected by Poon (1993) as more relevant to mature industries and less so to turbulent, dynamic ones such as tourism. A more positive view of Porter's reference has been taken by Leiper (1996 pers.comm.), who has argued that Porter classes industries into three categories: emerging, fragmented and mature. He maintains that because tourism is a collection of industries as opposed to a monolithic one, Porter's different industry characteristics may apply to different parts of tourism at different times. Tse and Olsen (1990) attempted to apply Porter's theories to the restaurant sector, but found the typology inadequate because of 'the fundamental differences' between the goods and services sectors. Poon's *Tourism, Technology and Competitive Strategies* (1993) has been the most comprehensive and convincing application of competitive strategy to tourism. Her many Caribbean examples offer similarities with a number of Pacific island resort destinations.

In examining the broad social, environment and economic context of island resort tourism, it is not the author's intention to test hypotheses of tourism competitiveness. The opportunity to examine some of the broader social and environmental issues essential to develop an overall comparative profile of tourism in the two regions is preferred to the more constraining 'auditing' approach. One of the authors of the recently developed Calgary model admitted that the applicability of the model to tourism regions as opposed to tourism nations was unclear at such an early stage in the model's development (Crouch pers.comm.). Consequently the author opted not to use the Calgary model as a yardstick to measure the relative competitiveness of the two destinations.

The present research examines the proposition that the regulatory and operational environments for Queensland and Fiji are now converging after quite different histories. The two regions are placed within the context of domestic and international (i.e. outbound from Australia) tourism. The present chapter places particular emphasis on the influence of air-inclusive holidays and the roles of governments and intermediaries. Comparing two sub-regions located within very different countries is complex but justified, since the field is neglected despite offering potential as a field of academic research. Are holidaymakers interested in regional concepts? As indicated in chapter 1, the tour operator Jetset has combined the whole South Pacific in a single *Resorts* brochure, suggesting that, for that company at least, countries are of little significance for marketing purposes and by implication sub-regions are less important still. On the other hand, increasing consumer interest in issues of authenticity going beyond national cultural stereotypes suggests that regional tourism would merit an enhanced role, emphasising unique regional attractions.

THE SCALE AND NATURE OF ISLAND RESORT OPERATIONS

Table 4.1 outlines the very different scale of operations in the Whitsundays and Mamanucas with the accommodation capacity of the Queensland group almost three times that of the Fiji group. It should be noted that some of the figures are approximate, either because of current resort expansion, or uncertainty over whether certain types of accommodation should be included in total holiday accommodation stock (the privately-owned apartments on Hamilton Island, for example, have been excluded).

Since tourism is an international phenomenon, a key structural issue for a tourism destination region is the extent of the linkages with international capital, labour and demand. The presence of transnational corporations can potentially provide access to all these resources for developing countries (Dunning and McQueen 1982a, 1982b). One might expect Fiji to be more dependent on overseas resort management. Britton (1983) equated the 'underdevelopment' of Fiji with the dominant role of international capital and management. Most of the Mamanuca resorts are managed by Caucasians on behalf of (predominantly) Australasian companies. Two exceptions

Table 4.1 Room capacity in the Mamanucas and the Whitsundays

Mamanuca resort	Room capacity	Whitsunday resort	Room capacity
Beachcomber	25	Brampton	108
Castaway	66	Club Med	200
Mana	138	Daydream	305
Matamanoa	20	Hamilton	667
Musket Cove	42	Hayman	225
Naitasi	38	Hook	12
Navini	9	Palm Bay	14
Plantation	110	Radisson	140
Tavarua	12	South Molle	202
Tokoriki	19		
Treasure	68		
Vomo	30		
TOTAL	577	TOTAL	1,873

Sources: RACV 1992 and TCSP 1992

are Mana Island which is managed by a Japanese company and Tavarua by a US-based company. Despite the dominant overseas role, only Vomo Island with its capacity of thirty guests is managed by a significant international corporation, ITT/Sheraton. In contrast, international management companies are now significant in the Whitsundays with Club Med, Daydream with Southern Pacific Hotel Corporation (SPHC), Holiday Inn (at Hamilton Island) and (until its recent takeover by an Australian corporation) Radisson Long Island.

The growth of transnational chains into the Whitsundays region was welcomed warmly by all general managers. Though the promotion of international chain brands may be less of an inducement to domestic travellers, many overseas visitors seek the reassurance of familiar names (Dunning and McQueen 1982a). There was an expectation that the standards of management would rise. In contrast to the Whitsundays, Fiji has struggled with low occupancies and international chains have been reluctant to invest or manage in Fiji. The small Sheraton Vomo Resort is a recent exception.

Travel agents and tour operators assist the process of destination image building. Travel agents are critical to the consumption process through their participation in almost all all-inclusive holiday sales (as indicated in table 4.3) and because of their power to persuade consumers face-to-face of the merits of particular tourism products. Tour operators 'create the package' by assembling the various components such as airline seats, airport transfers and hotel beds and 'packaging' them into an easily consumable product. Collectively agents and operators can induce demand provided that they have access to competitive prices. If agents and operators become unfavourably disposed to a country such as Fiji for reasons such as excessive price increases by airlines or resorts, a perception that the range or quality of product available is inadequate, or political instability, they have the power to redirect bookings to Bali, Hawaii or alternative destinations. The power exerted by metropolitan-based distributors of travel is a key dilemma for countries like Fiji, determined to pursue an independent course in the post-colonial era.

In the present research, an air-inclusive holiday is taken to mean leisure-motivated travel undertaken by Australians where air travel and at least one night's accommodation are purchased from a single source. This differs from European and North American definitions which usually incorporate all transport, accommodation, meals and transfers (Middleton 1989) and reflects the less structured and more independent character of Australian travel behaviour. The Chairman of the Australian Council of Tour Wholesalers (ACTW) indicated that operators cannot apply a minimum stay requirement because of the different conditions imposed on 'inclusive tour' fares by different destinations and that his organisation does not adhere to any (T. Milmore pers.comm.). The Chief Executive of Jetset

Tours went so far as to describe the term inclusive tour as 'largely irrelevant' in Australia on the basis that increasingly flexible structures make them indistinguishable from 'point-to-point' travel (i.e. air only) (D. Clarke pers. comm.). In the present research the term 'inclusive holiday' is preferred.

INTEGRATION IN THE AIR-INCLUSIVE HOLIDAY SECTOR

Poon (1993) has proposed some key factors to characterise the structure of the contemporary tourism industry and advises firms and destinations to adapt to ensure their future competitiveness. The factors are: the concept of an 'industry value chain', the value of 'diagonal integration', deregulation, an increased emphasis on product 'quality' and the need to be innovative. She maintains that the last two factors are a necessary response to the decline of the 'old tourism' and the rise of the 'new tourism'. In contrasting the 'old tourism' and the 'new tourism' Poon depicts the former as practising 'mass marketing, standardised, limited choice and inflexible holidays' and the latter as 'greener, more individual, flexible and segmented' (1993: 18). The false dichotomy that old is bad and new is good which appears to underpin the analysis is a limitation of Poon's work which appears to exaggerate certain contrasts. Nevertheless, her observations highlight some alternative responses for the Whitsundays and the Mamanucas as a means of changing or moulding consumer tastes. Poon has sought to explain all of the elements responsible for bringing about the end of the old and the beginning of the new and in the present research her findings are used to illuminate the key features of Whitsundays and Mamanucas tourism.

Her expression 'industry value chain' is a refinement of the long-established marketing principle that consumers and suppliers are connected by a chain of distribution. Instead of assuming equality between all elements of the chain, the industry value chain asserts that some elements enjoy a stronger bargaining position than others and that this relative influence may change over time. One example cited by Poon is as follows: '. . . tour operators removed from hotels the worry of whether their bed-nights would be sold or not. Hotels thus had limited control over who their clients were, where they came from or the marketing of their products' (Poon 1993: 233). The applicability of this reference to the Australian market is somewhat limited by the fact that, unlike what occurs in Europe, Australian tour operators do not actually commit themselves to bed-nights – their role is confined to marketing them. This is an important distinction between distribution channels as they apply in Australia and in the rest of the world. Poon states that the improvement of yield management systems, direct marketing and the prospect of repeat visitation has lessened the need for resorts to allocate blocks of rooms to tour operators at heavily discounted rates. The present research asks what (if any) changes have occurred and may be expected to occur in the value chain affecting the Whitsundays and the Mamanucas?

'Diagonal integration' occurs when a business forms an association with other businesses to be able to offer a range of complementary products to an identified group of consumers. The Australian Fly Buys scheme, for example, offers consumers free flights in return for purchases made through selected outlets including department stores, petrol stations, banks and restaurants. Collectively these outlets offer a single product (Fly Buys) with a view to increasing customer patronage. The exchange of customer databases is a key element of Fly Buys and often a characteristic of diagonal integration. Loyalty Pacific who manage the scheme claim to have enrolled 1.25 million members during the first six months of operation (Elliott 1995). Despite such rhetorical (and promotional) claims, no evidence is currently available concerning the impact of Fly Buys on the turnover or margins of participating businesses. Poon distinguishes diagonal integration from the longer established expressions vertical and horizontal integration.

By industry deregulation, Poon (1993) refers primarily to the *Airline Deregulation Act* 1978 (US), but deregulation affects both the Whitsundays and the Mamanucas. Poon again draws primarily from US examples in highlighting the pressure by consumers for better 'quality' products. She states that 'in the late 1970s, only 40 per cent of US buyers ranked quality as being at least as important as price in their decisions, but in the 1990s the figure is over 80 per cent and rising' (1993: 256). She asserts a growing demand for products which promise and deliver good quality at mid-range prices. Finally Poon's assertion that firms need to innovate highlights the danger of being seen as a 'me too' destination which offers little different from what is available elsewhere. Collectively, Poon's assertions offer a useful taxonomy for assessing the effectiveness of the Whitsundays and Mamanucas as destinations.

In the past the domestic and outbound air-inclusive holiday industries in Australia have been quite separate. The various companies involved in the packaging of overseas destinations have declined to involve themselves in domestic activity. Domestic tour operators (including the airlines and state tourism commissions) have generally confined their activities to travel to and within Australia. Entirely different companies have packaged sun, sea and sand domestic destinations and equivalent international destinations. These may be defined as destinations within a four- to seven-hour flying time from Melbourne and Sydney. Although the operators are different, the fundamental elements of the domestic and international 'products' are very similar. The distribution channels used for domestic and international holidays were also different. Traditionally retail travel agents in Australia have shown little interest in domestic travel, with a much higher proportion of domestic sales handled directly by the airlines as illustrated in table 4.3.

Domestic air-inclusive holidays are dominated by Ansett Australia and Qantas Airlines. The state government travel bureaux own and operate their own holiday brands, most notably Sunlover Holidays in Queensland. Other operators are either highly specialised (for example, ecotourism and adventure operators), or concentrate on other transport modes such as coach (for example, Australian Pacific Tours). Neither of the latter categories is a significant source of business for island resorts. The airlines achieved domination, integrating vertically through the purchase of hotels, resorts, inter-island launches and coach operations. This contrasts with developing country destinations such as Fiji where the national airline lacked the resources to invest in complementary industry sectors. In Fiji, it was Qantas which pursued vertical integration by purchasing a controlling share in an inbound operator, Sun Tours (a share now divested).

Airlines are involved in the ownership of Australia's two largest outbound operators, totally in the case of Jetabout (100 per cent owned by Qantas) and partially in the case of Jetset (50 per cent owned by Air New Zealand). The airlines also enjoy close commercial arrangements with most of the other tour operators since the latter are dependent upon the supply of airline seats on scheduled flights. Qantas Holidays, through its Jetabout and Viva! brands, is the largest tour operating block to Fiji and was cited as the most significant by most resort managers in the Mamanuca islands. Fiji is packaged by approximately thirty other tour operators out of Australia, many of which are not airline owned. Overall those tour operators free of airline ownership have offered more significant and more direct competition on international routes such as to Fiji than their domestic equivalents, despite the strong control exerted by airlines over the supply of seats through the purchase of Viva! by Qantas and the collapse of Venture Holidays' Victorian division, undermining the influence of independent tour operators. Lacking its own tour operating arm, Air Pacific has been less able to manipulate demand than Ansett and Qantas.

The airline and tour operating environments have been strongly influenced by government regulations. In Fiji, the environment is determined by the relevant Air Services Agreement (ASA) between the Governments of Australia and Fiji. The regulatory framework was traditionally described as the 'Two Airline Agreement', an arrangement legislated through a series of *Air Navigation Acts 1920–1973* (Cwlth). The Agreement consciously sought and achieved relative equality (or duopoly) between the government-owned Australian Airlines (until its merger with Qantas in 1992) and the privately-owned Ansett Australia, with each airline controlling approximately half of the total business. Domestic competition was progressively encouraged and regulations removed through the 1980s culminating in the formal beginning of airline deregulation in November 1991. The domestic regulatory framework is now much less prescriptive.

Liberalisation between Australia and Fiji has taken a different form. The change is partially explained by the relationship between Qantas and Air Pacific and partly by government policy. The Australian Government sold a 25 per cent equity share of Qantas to British Airways as the first stage in a process of privatisation and a public float (which took place in 1995). The impact of impending privatisation on Qantas over the period 1992–5 was twofold. Firstly the international destinations served by Qantas began to compete directly with its new domestic assets (it inherited five Queensland island resorts from Australian Airlines). Secondly it re-evaluated traditional destinations such as Fiji in the light of the increasingly global ambitions that it shared with partner British Airways. This strategy has involved the shedding of 'non-core' routes. Whilst there is little prospect of total withdrawal from Fiji, the destination may well become marginalised.

Qantas has taken an increasingly 'hands-off' approach to Fiji and ceased its destination marketing activity prior to its merger with Australian Airlines. Since 1985 Qantas has provided Air Pacific with a 'corporate support' role, initially as a means of getting the smaller carrier back on its feet after a period of serious losses and Air Pacific has acted increasingly independently of Qantas as its financial position has improved. The result has been an increasing share of capacity for Air Pacific (still 80 per cent owned by the Fiji Government) and (arguably) a reduction in competition. In contrast, Bali, a key island resort competitor is currently serviced by three carriers – Qantas, Ansett and Garuda Indonesia. In Fiji there is a perception amongst the tourism industry (D. Smith pers.comm.) that Air Pacific is excessively influenced by Qantas and that the Fiji Government delegates most of its aviation negotiations to Air Pacific because of the inadequate resourcing of the Fiji Ministry of Aviation (consisting of three staff at the time of writing). Whilst Air Pacific claims to be 'profit driven', its negotiating role is virtually indistinguishable from that of the Fiji Government, thereby imbuing apparently commercial decisions with a strong bureaucratic and political dimension. In the case of the Whitsundays, competition is still impeded by Ansett's monopoly on the use of Hamilton Island Airport (despite denials by airline representatives). Qantas is at least able to compete (albeit from a position of disadvantage) through pursuing a strategy of bolstering its activities via Proserpine Airport. In contrast competition on the Australia–Fiji routes has probably lessened.

Prior to the development of long-haul Boeing 747 aircraft capable of crossing the Pacific non-stop, Fiji played a significant role as a refuelling stop for flights between North America, Australia and New Zealand. Most flights now overfly Fiji, reducing the significance of the country in the region and in consequence, the negotiating power of the national airline. The increasing internationalisation of aviation, characterised by cross-national alliances, has further marginalised Air Pacific. The latter does not enjoy the range of significant international affiliations enjoyed by Ansett and Qantas, nor does it have a substantial home industry or population base which would allow a domestic network to develop. Through the 1980s all of the North America-based carriers (Canadian Airlines International, Continental Airlines and Pan American Airlines to name but three) withdrew their services from Fiji. A number of European and Asian carriers did the same (UTA, Air India and Japan Airlines). This made the country entirely dependent on airline capacity offered by airlines operating out of the South-West Pacific (Qantas, Air New Zealand and Air Pacific) and the various smaller regional carriers. Fiji's strategic position as a stopover destination had previously shielded it from the uncomfortable reality that confronts many small island destinations. Since the domestic demand for international travel in such countries is small, traffic is often predominantly derived from a single overseas source. This makes the economics of airline operation less attractive in island destinations. In the absence of significant stopover traffic, their role is typically confined to short-haul single destination holiday travel. This is characterised by small seat capacity and excessive ticket prices. The dilemma for Fiji is exacerbated because Australia, its primary source of business, is sparsely populated relative to the USA and Europe. In Fiji's case it is harder to justify charter flights which would otherwise offer the prospect of better economies of scale, more flexibility for tour operators and cheaper prices. The withdrawal of major airlines was a significant blow to tourism in Fiji.

Ansett has no involvement and little apparent interest in flying from Australia to Fiji. This reluctance may benefit Air Pacific's profitability but may disadvantage Fiji in a geopolitical sense. Though the possibility of Ansett operating flights from Australia to Hawaii via Fiji has been canvassed (Marshman 1994), the prospect of further expansion between Australia and Asia is more likely. This suggests that outlets such as travel agents which have no preferential commercial association with Qantas will be less likely to favour Fiji as a destination. The power of agents to influence demand is exemplified in the following quotation from Poon (1993) who states that 'a travel agent could route a passenger on a flight with an inconvenient stop in order to gain an extra commission on a preferred carrier, rather than provide the client with a more convenient flight' (1993: 232). One example of disadvantage concerned airline frequent flier programmes. Air Pacific neither had a scheme of its own, nor was a member of any other scheme. Whilst frequent fliers are likely to earn their qualifying air miles and to redeem them through travel to domestic destinations such as the Whitsundays, only Qantas frequent fliers can do so in the case of Fiji. The (dominant) Air Pacific seats could not be used for either purpose. In apparent recognition of this competitive disadvantage, in 1996 Air Pacific became a partner to Qantas' Frequent Flyer Program.

In table 4.2 some of the key affiliations and commercial agreements relating to Ansett and Qantas are listed. They are symptomatic of the importance ascribed to international networks and alliances by the airlines.

The wide range of Queensland island resorts owned by the two airlines is a notable feature of the table. Both airlines are interested in boosting sales to these islands, though Ansett's interests are more concentrated in the Whitsundays than are Qantas' interests. Qantas did recently promote one of the Mamanuca islands, Castaway, for the first time in its November/December 1994 *Frequent Flier Newsletter*, using Qantas Jetabout as the tour operator. Significantly, though, travel was restricted to Qantas flights and the emphasis given to the airline's five Queensland island properties in the same newsletter was much greater.

Since Qantas and Australian Airlines were merged, the former holiday programmes of the two airlines have been administered together out of Sydney, though until recently the holiday division has maintained separate

Table 4.2 Travel industry alignments and affiliations with Ansett and Qantas

	Ansett Australia	*Qantas*
Accommodation	Hamilton Island Hayman Island Accor Asia Pacific ANA Hotels Holiday Inn	Bedarra Island Brampton Island Dunk Island Great Keppel Island Lizard Island Intercontinental Rydges Radisson Country Comfort
Car rental	Avis	Hertz
Tour wholesalers	Ansett Holidays Jetset	Qantas Jetabout Viva! Creative Tours Sunlover Holidays
Retail travel	Traveland Jetset Travel Metro Travel	American Express Harvey World Travel UTAG Travelstrength National Travel
Airlines	Cathay Pacific Singapore Airlines United Airlines Malaysia Airlines Swissair Austrian Airlines All Nippon Airways	British Airways US Air Scandinavian Airlines Canadian Airlines American Airlines
Credit cards	Diners Club	American Express

Source: Ansett and Qantas Frequent Flier Bulletins plus data from major retail outlets, October/November 1994
Note: The relationships noted above are rapidly changing

brand identities under the labels Viva!, Jetabout, Creative Tours and Qantas Australian Holidays. The close alignment of major tour wholesalers Jetabout and Viva! to Qantas should be of concern for Fiji, given Qantas' declining interest relative to its Queensland resorts. The separate identities of Jetabout and Viva! Holidays were merged in mid-1996 to trade under the new brand, Qantas Holidays. This move merely confirms what was already widely known – the various Qantas brands were in practice the same thing.

Within Australia, some competition to the airline holiday packages is provided by small independent wholesalers. A notable example is Australian Reservation Centre ('1– Call') which has commenced holiday packaging in direct competition with Sunlover and the airlines to Queensland, including the Whitsundays.

REGULATION AND THE AIR-INCLUSIVE HOLIDAY SECTOR

Prior to deregulation in 1991, domestic aviation and domestic air-based packaging was a virtual duopoly. Ansett's post-deregulation advertising proclaimed that deregulation allowed the airline to be 'What it had always wanted to be' but the highly regulated environment provided Ansett with stability and guaranteed profits. It undoubtedly resulted in higher prices for domestic holidays and consequently with less intense domestic competition. Of the two airlines, it was the better financed Ansett which took the risk of developing the Hamilton Island airport in partnership with resort owner and developer Keith Williams. In consequence, tourism in the Whitsundays has favoured Ansett. Ironically the development of Hamilton Island Airport took place within the highly regulated pre-deregulation environment. Ansett enjoyed a larger domestic capacity and market share than Qantas until the 1990s when Qantas pulled substantially ahead, boosted by on-carriage of its international passengers.

The Qantas/Australian Airlines merger lessened some domestic and international contrasts. Fiji's location within the south-west Pacific also comes into play. When launching domestic airline deregulation, the Commonwealth announced the phasing in of an 'open skies' policy throughout Australia and New Zealand as part of the Closer Economic Relationship (CER) between the two countries (Hall 1994a). Progress towards this objective within Australia has been slower than originally planned (New Zealand substantially deregulated its domestic operations some time ago), with the Australian Government reluctant to make concessions to Air New Zealand on domestic routes. Were Air New Zealand to assume certain domestic routes there could be competitive repercussions for the Whitsundays, though the Ansett monopoly at Hamilton Island Airport currently seems non-negotiable. Since Fiji falls outside the sphere of CER, the impact on competition between Australia and Fiji is not immediately apparent. Fiji does benefit from trade integration with Australia and New Zealand through the South Pacific Area Regional Trade and Co-operation Agreement (SPARTECA) and Air Pacific has engaged in commercial arrangements with Qantas and Air New Zealand to enable it to obtain favourable leases on aircraft. As it stands, however, Fiji's aviation industry cannot expect to benefit directly from CER. As an outsider to what is in certain respects a trade block, Fiji may be in danger of peripheralisation. Just as Australia has feared marginalisation in a world increasingly dominated by trade blocks such as the North American Free Trade Association (NAFTA) and the European Union (EU), Fiji is threatened by marginalisation within the south-west Pacific. In late 1994, the Australian Government announced its intention not to proceed with the opening of domestic air routes to Air New Zealand. This stalled the rapid progress of Australasian deregulation and provides Fiji with a temporary reprieve from ever more intense competition from Australian domestic destinations.

Because of the régime of individual air services agreements (ASAs) between each of Australia, Fiji and New Zealand, Pacific routes have thus far been unaffected by the more liberal approach with competition between Fiji and Australia limited to two carriers closely associated through 'code sharing'. The potential of code sharing to reduce competition was evident between Australia and New Zealand. Over a two-year period ending in 1994, most flights across the Tasman were 'code shared', meaning that those booking with either Qantas or Air New Zealand would typically be unaware of which airline would operate the actual sector booked. Code sharing served the short-term interests of the airlines but was finally abandoned when the anti-competitive consequences of the practice were alluded to by the Trades Practices Commission. Through the period of code sharing, Qantas held a 19.9 per cent equity share in Air New Zealand. As previously mentioned, the practice continues between Australia and Fiji with Qantas owning a 10 per cent shareholding in Air Pacific. Air Pacific's dependence on Qantas is explicit in Melbourne where the Fiji carrier has no office and relies on bookings made through Qantas which 'represents Air Pacific's interests'.

Governments' pursuit of 'national interest' in aviation at the expense of free trade has been the subject of growing criticism. The Managing Director of Singapore Airlines has argued that the benefits of deregulation are best served in a multilateral trading environment (Kong 1994). There is currently little prospect of such a development within the South Pacific. The contrast between the liberal Australian environment and the more rigid environment in the South Pacific precludes any real 'convergence' of civil aviation activity between Australian gateways and between Australia and Fiji. It is somewhat ironic that the diminishing interests of Qantas in Fiji has resulted in enhanced influence for its 'competitor' Air Pacific as evidenced in the market share data presented in the previous chapter.

Qantas is also the minor player in the Whitsundays because of its inability to access Hamilton Island Airport. The difference is that whilst Qantas has little apparent interest in enhancing its capacity to Fiji, it would greatly like to enhance its holiday capacity (through Hamilton Island) to the Whitsundays.

If Air New Zealand were to be granted 'Fifth Freedom Rights' (the right to carry passengers from a second country to a third country such as Fiji), it would undoubtedly enhance the competitive environment. It is regarded as a competitive 'niche' or specialist carrier within the south-west Pacific. The strict regulatory environment in place between Australia and Fiji prevents it from increasing competition on the relevant routes. One might argue that Air New Zealand would provide stronger competition to Air Pacific than Qantas because of its stronger focus and strategic commitment to the South Pacific region (the South Pacific countries are relatively more significant to New Zealand than to the much larger Australia). At the time of writing, Ansett and Air New Zealand are engaging in an increasingly close relationship with the purchase of a 50 per cent equity share of Ansett by Air New Zealand. This opens up the possibility for Air New Zealand to bid for capacity on flights between Australia and Fiji in conjunction with Ansett.

MARKET SIZE AND MAIN DESTINATIONS

To understand the positioning of the Whitsundays and the Mamanucas we need to look at the relative scale of domestic and international holiday travel. Domestic travel in Australia accounted for some 48.3 million trips in the financial year 1993–4 (including travel for business purposes), with about 5.0 million of these by air, amounting to 11 per cent of total domestic trips (Bureau of Tourism Research 1994). There were 2.27 million short-term resident departures in 1993–4, a figure forecast to rise to 3.48 million by 1998–9 (Tourism Forecasting Council 1995). Precise figures for inclusive holidays are difficult to ascertain. In preparation of this chapter the author contacted all of Australia's top thirty tour operators but obtained only minimal data about turnover, profitability and market share. No interfirm comparison exists in Australia for the tour operator sector. The retail travel sector is slightly better served by research commissioned by the Australian Federation of Travel Agents (AFTA).

According to Jetset Tours, fully inclusive holidays declined from about 25 per cent of total sales in 1986 to between 10 per cent and 15 per cent in 1990 and are now an 'outmoded form', largely superseded by modular format travel whereby the consumer can combine a number of discrete elements (D. Clarke pers.comm.). He claimed that of all outbound destination regions, the fully structured packages to the Pacific islands sold by Jetset have the shortest duration and the lowest spend per customer on average.

Comparing airline-owned tour operators with independent tour operators, the latter are able to strike a balance between those airlines which offer them the largest margin (either through overrides or net fares) and those which will be most sought after by consumers. Generally the bigger the operator and the more seats that it can guarantee to an airline, the better the margin that can be negotiated. This situation also affects the use of hotels.

With the limited Australian population of approximately 18 million, the absence of economies of scale is a dilemma for independent tour operators. The bulk of consumer spending goes to the airlines and most operate their own wholesaling divisions. The small to medium-sized resorts found in the Mamanucas offer tour operators small sales volumes. Unless such small resorts are seen as offering customers something quite distinct, they will be bypassed by operators in favour of large properties, especially if they are perceived as mainstream sun, sea and sand destinations differentiated predominantly by price. Such economics work in favour of a single destination (for example, Hamilton Island) which can offer substantial capacity compared to a range of properties under separate management offering limited capacity (the twelve Mamanuca resorts with their total capacity of less than 600 rooms). Managers in both the Whitsundays and the Mamanucas pointed to pressure from tour operators to make an ever larger proportion of their products 'commissionable'. Historically the resorts and airlines provide tour operators with commission on airfares, transfers, accommodation and (sometimes) breakfast. Resort managers argue that pressure to make lunches and dinners commissionable (that is, part of the pre-purchased package) squeezes the profit margins unacceptably, especially as they are under constant pressure to reduce their rack rates. The squeeze on tour operator margins suggests that there are too many players and a shake-out is due.

Until very recently, Air Pacific has opted to work through third parties in preference to establishing its own holiday division, thus prompting many independent wholesalers to feature Fiji holidays out of Australia. This contrasts with the airline package dominated Whitsundays. In the type of environment exemplified by Fiji, wholesalers offer the airlines access to wider distribution channels and a brand name with which customers can identify, for example, Jetset and Jetabout. This may be important where the airline itself (that is, Air Pacific) is little known by consumers. The longer-term dilemma for the airline is that an over-reliance on tour operators exacerbates the already low profile of the airline brand in Australia. This is an issue now being addressed by Air Pacific in its increasing spending on brand awareness marketing (E. Dutta pers.comm.). The structure of the international tourism industry (where control of the distribution network within source markets is vital) mitigates against small, resource-starved entities such as Air Pacific, based in developing countries, from gaining the necessary profile within metropolitan countries. The Mamanucas themselves have become over-dependent on overseas (i.e. Australian) based tour operators for selling their product and, importantly, for cultivating relations with key travel agents. If, as Poon (1993) has argued, tour operators are the weakest and most vulnerable element of the chain of distribution, this is a real threat.

The domestic holiday division of Qantas (then Australian Airlines) reported (King 1991) that in gross terms 75 per cent of all domestic air package sales are to Queensland, made up of 35 per cent to the Gold Coast and the remaining 40 per cent to the Queensland Islands and Far North Queensland. Tasmania accounts for 10 per cent of tour packages. It was reported that accommodation-inclusive packages accounted for only 7.5 per cent of total passenger sales on domestic routes. This was less than a quarter of the number of seat-only leisure travel sales (31.5 per cent). Recent pronouncements by Qantas suggest that short-break packages are growing fastest.

DOMESTIC AND INTERNATIONAL AIR-INCLUSIVE HOLIDAYS: A COMPARISON

Can Fiji and Queensland island resorts be characterised in terms of the 'old' and the 'new' tourism? The type of air inclusive tourism characteristic of the Gold Coast seems to parallel the 'old-style' European package holiday experience where holidaymakers 'migrate' from cooler North Europe climes to the urbanised, high-rise (largely) Spanish resorts such as Benidorm, Palma and Torremolinos. Whilst Australian domestic holidays are based on scheduled flights, most European resort holidays are based on charter flights. Secondly the climatic differences are less extreme between the generating and receiving regions. Australians do not need to travel overseas for a 'sun, sea and sand' holiday experience and the Gold Coast and the Sunshine Coast are within easy driving distance from Sydney (less so from Melbourne). Because of their location in north as opposed to south-east Queensland, the Whitsundays are much less accessible by car. Except for travellers with extended holiday periods, air-based travel is the only real option for holidaymakers. The closest parallel between the Whitsundays and Surfers Paradise is Hamilton Island which has been described as 'Surfers on an Island'. Surfers Paradise itself is large enough to have become an archetype, but remains an exception within Australia. In many respects Australia can be regarded as an early exponent of the 'new tourism', having broken with the traditional tour format early in its development (King and McVey 1996). The non-availability of charter flights has probably reinforced this trend.

The promotional calendar of domestic air-inclusive holidays is dominated by the annual launch of holiday brochures by Ansett and Qantas. Queensland is the major holiday destination for both airlines. In 1994, Ansett produced four main full-colour brochures on Queensland Islands, the Gold Coast, Cairns and the Queensland Coast. In the case of Qantas, Queensland accounts for five of its nine Australian brochures. These are divided along regional lines. As is the case with Ansett, there are brochures for Cairns (identified as the 'Tropical North'), the Gold Coast, and the Queensland Islands. The other two might create some confusion for the Whitsundays. The coastal areas of the Whitsundays are grouped together with the Sunshine Coast, whilst Mackay is grouped with Townsville and Rockhampton. Consequently a consumer wanting information about Mackay, the 'gateway to the Whitsundays', would need to read one brochure, someone wanting a Whitsunday Islands holiday would need to refer to the Queensland Islands brochure and someone considering a holiday at Airlie Beach would need to refer to the Whitsundays/Sunshine Coast brochure. It is worth noting that Sunlover Holidays offered fourteen Queensland brochures in 1994. The two relevant to the Whitsundays are 'Queensland's Islands' and 'Whitsundays and Mackay'. Significantly, Sunlover have chosen to group the Whitsundays together with the nearby Mackay rather than with the more distant Townsville. According to the Product Development Manager of Sunlover (S. Brewster pers.comm.), the groupings are determined by the potential sales volume which is expected to cover the cost of brochure production consistent with the increasing product-by-product financial accountability referred to in chapter 3. He did, however, acknowledge that political issues can also intervene.

In Europe, holiday brands such as Thomson Holiday products dominate destination marketing and reduce the importance of the actual name of the destination through offering generic sun, sea and sand holidays under their own brand. This type of promotion has resulted in tourism labels such as *Costa del Sol* to take increased prominence over the destination of Spain. In contrast a reading of the Australian holiday region brochures suggests that geography continues to hold sway. Marketing the geographical or administrative entity is the norm, not the operator brand. Because of the wide range of products on offer, individual regions within Queensland warrant their own brochures, unlike the other states which are accounted for by a maximum of one brochure each by Ansett and Qantas Australian Holidays. The name Whitsundays has the advantage of strong consumer appeal, whilst being an accurate description of the region itself (that is to say, it is not contrived). In contrast, the name Mamanucas shares the advantage of geographical correctness, but it lacks consumer awareness. Altering names by enforcing artificial labels such as 'Gold Coast' and 'Sunshine Coast' reduces perceived authenticity, though it has the potential to enhance consumer recall.

Geographical accuracy gives way to the generic 'resorts' label in the case of brochures featuring the South Pacific. Australia's largest wholesaler of international travel, Jetset, uses the label 'resorts' to describe its brochure consisting of the destinations Fiji, Hawaii, New Caledonia and Vanuatu. The other brochures in the Jetset range typically use the name of the actual destination, for example, 'Singapore', 'Americas', 'Singapore', 'Hong Kong' and 'Thailand'. For Jetset, the South Pacific owes its significance to its resorts. Accommodation and resort options occupy all but one page of the company's main Pacific brochure (the one page concerns 'cruises and sightseeing'). The tendency of major tour operators to depict the South Pacific as no more than a series of resorts is commonplace. In 1996 Jetset dropped its South Pacific Islands programme entirely.

Despite strong airline control over the product, direct selling of package holidays in Australia is very limited. The former Australian Airlines sold approximately 50 per cent of its point-to-point air tickets directly to the public. In the case of package holidays the proportion is about 15 per cent. The reason that travel agents are interested in inclusive holidays is pecuniary. Package tours pay the agent 10 per cent, point-to-point domestic tickets 5 per cent (officially). To the extent that most consumers of domestic package holidays purchase at the same travel agencies where outbound tours are sold, there is direct competition between the two types of destination. Table 4.3 shows the greater importance to the travel agent of selling domestic holidays as opposed to point-to-point air tickets.

Both Ansett and Qantas have departments dedicated to inclusive holiday products. These departments are responsible for product development and for distribution through the travel agent network. Unlike in Europe where most package prices are inclusive, Australian brochures usually separate the accommodation and air components. Since air-inclusive package tours in Australia use scheduled and not chartered aircraft, the airfares available are

Table 4.3 Estimated growth and share of the Australian travel market 1990–2000

	1990		1995		2000	
	Outbound and domestic ($000)	Travel Agency share (%)	Outbound and domestic ($000)	Travel Agency share (%)	Outbound and domestic ($000)	Travel Agency share (%)
International airlines	2,800	85	3,400	75	4,600	75
Domestic airlines	2,100	55	3,100	60	4,400	55
Wholesale packages (excl. flights)	725	85	905	85	1,200	85
Ad hoc accommodation	1,500	20	1,900	20	2,260	20
Travellers cheques	1,800	30	2,200	35	2,900	40
All other expenditure	460	20	540	25	700	25
Total	9,385	54	12,045	55	16,060	54

Source: Access Research 1992
Notes: Estimates growth over 1990 base in constant dollars. Includes business travel.

diverse. The airfare discounts applying to the main annual airline holiday brochures range from 25 to 45 per cent depending on the level of restriction applying to the relevant ticket. More recently the emergence of 'super discount' brochures, usually with limited duration off-season applicability, have introduced discounts of up to 70 per cent, though the all-inclusive nature of the price precludes consumers from knowing the true nature of the discount. In the case of the main annual brochures, the airfare is typically not specified and determination of the appropriate fare is left to the travel agent. Holidays during school holiday periods are soon sold out, limiting the accessibility of air-inclusive holidays for less affluent families.

The relatively high prices charged for standard (that is, non-discounted) domestic packages was explained by the limited volume and wide product choice. Even relatively low demand destinations feature a diverse range of accommodation and room configurations. A positive dimension is that Australian domestic packages are less typical of the 'old' tourism than their European counterparts. The higher cost involved in offering the consumer a wide range of alternatives is no longer seen as a disadvantage. Information technology and guaranteed bookings (for example, credit card confirmations by telephone), are making economies of scale less essential.

In Europe whilst standard holiday packages can be obtained cheaply, consumers who make non-standardised requests must pay much higher prices. Mass production holds sway. In Australia, there has been no penalty payable for making extraordinary demands and consumers have had little incentive to buy a standard package. Traditionally European package tour products have been pushed vigorously through the retail network using attractive lead prices and heavy seasonal and standby discounts whilst Australian package products have been characterised by high cost and low volume. Since deregulation, however, the marketing of air-inclusive holidays has taken on many of the features of European-style commodity travel promotion, with a heavy emphasis on so-called 'price leaders' and less lavish two-colour leaflets as opposed to full-colour catalogues. The range of choice available through 'flyers' is narrower than what is offered in the full-colour annual brochures. By 'flyer' we refer to a two- or three-colour leaflet or promotional item which features discount product for a limited time. Widely used by tourism industry personnel, the term is not to be confused with 'brochure' – a main booklet with extended applicability.

Sunlover Holidays which offer both surface transport-based holidays and air-based holidays to Queensland claimed to sell 80,000–100,000 packages per year, of which 30 per cent are air-inclusive holidays. Qantas claimed to have sold about 500,000 domestic packages in 1993, of which 86,000 were to the Gold Coast. The company was unwilling to indicate the percentage accounted for by specially discounted product under the *Getaway Deals* label but pointed to a 96.5 per cent increase in this category over 1992, suggesting a changed composition of sales (Gladstone 1995). The growing proportion of sales accounted for by discounted product began with attempts to revive consumer confidence after the disastrous effect of the pilots' dispute which curtailed domestic air services for several months.

During the 1970s and 1980s the Federal Government had allowed Australian Airlines (then known as TAA) to purchase five of the Queensland islands with a view to enhancing Queensland's tourism resort infrastructure. These islands were owned and operated by a division of the airline called Australian Resorts. The high costs associated with operating in an isolated island environment was already causing Australian Resorts to lose money. The pilots' dispute reinforced this trend. In the aftermath of the dispute, Australian Airlines started its holiday discounting in the area over which it had greatest control – the island resorts. Discounts of between 40 per cent and 56 per cent became the norm and most of the product sold in 1990 was discounted. If Australian Airlines hoped to quarantine the effects of its island discounting it was to be disappointed. Mainland Queensland could not afford to be left out. Before long Ansett and Australian were discounting holiday products to all of the major holiday destinations including the Gold Coast. The Queensland islands were increasingly competitive with Fiji from 1990. Once Qantas took responsibility for the marketing of these holidays, the competition became more obvious.

In a bid to stimulate demand, the minimum stay required for travellers to qualify for tour-basing fares was reduced from seven to five nights. Weekend packages to the major cities including Cairns were introduced with a minimum stay of one Saturday night. This increasing flexibility has diluted the previously accepted definition of a package tour. An oversupply of hotel beds and air seats throughout Australia enforced the need for flexibility and this more flexible arrangement now appears well established. A recent entrant to the Queensland package holiday business, 1-Call Holidays, highlights those resorts featured in its brochure which offer one-night stays. Of the seven Whitsunday Island resorts featured in its brochure *Queensland Islands – Whitsundays and Capricorn Coast*, only one (Club Crocodile Long Island which requires a two-night minimum stay) is not highlighted. The shift from what used to be relatively inflexible (seven-night stays) to absolute flexibility (one night) signified the transformation of package structures.

As the airlines became increasingly preoccupied with yield, the sacrosanct nature of the traditional holiday package (seven nights plus) has been challenged. As long as there was little risk of yield dilution (higher-price passengers transferring to lower fares), the airlines were willing to eliminate some of the rigidities such as minimum stay requirements. The advent of deep discounting revolutionised Australian air-inclusive holidays. The discounted product was based on an attractive lead-in price, including airfare. Brochures were introduced using a two- or three-colour format and with a short lifespan (for instance six weeks to two months). The discounted product required simple booking procedures, limited consumer choice and heavy promotion through the retail network. Accommodation listings were limited and highly discounted. The production of flyers is common practice amongst most holiday packages to both the Whitsundays and to Fiji, indicating the convergence of marketing techniques.

The continuation by Qantas and Ansett of their traditional brochure ranges prompts a question about future viability. If most consumers are buying heavily discounted product based on low-cost leaflet style brochures, why bother with a high-cost glossy brochure range featuring destinations with low volume and (relatively) high prices? Should separate winter or other seasonal products be introduced thereby ending the era of fixed price year-long products? The answer provided by some of the smaller operators featuring Fiji has been to dispense with full colour brochures entirely and to rely on a series of two- or three-colour flyers. Unfortunately for Fiji, the absence of airline commitment to full-colour brochures may have tarnished the image of the country through the dissemination of low-cost promotional materials. Remarks (albeit isolated) made by focus group participants during the present study pointed to a minority perception that Fiji is a 'down-market' destination. Given the lessening compulsion to produce full-gloss annual brochures, it is possible that magazine and lifestyle catalogues may be introduced as they have been in Europe (Thomson Holidays offers a programme called 'Freestyle').

EMERGING STRUCTURES AND INDUSTRY TRENDS

Australian-based international tour wholesalers such as Venture Holidays and Prima Holidays might have been expected to operate domestic holidays. Such operators would require a margin (the difference between the price they pay and the price they charge) of 20–5 per cent off the air ticket price to allow them sufficient to offer 10 per cent to travel agents off the already discounted price. The airlines have offered 25 per cent off the standard airfare but not off the already discounted product. In these circumstances potential operators know that they would always be undercut by the airlines themselves. So why bother? Unless the number of independent tour wholesalers featuring domestic air-based holidays increases, the airlines will continue to develop high-cost, glossy, destination-based materials, though an increasing proportion of their sales will be generated by short-term lower cost promotions. A similar balance between leaflets and brochures will be evident in Fiji, though with a single airline (Qantas) in a position to dictate terms.

Before its demise, Compass Airlines (in which the Queensland Government owned a shareholding) began to engage in discount packaging through third parties in preference to establishing its own holiday division.

Had Compass survived, the role of domestic airlines in packaging might have changed. Though this did not eventuate, Compass did raise consumer expectations about the availability of deep discount fares and packages, thereby giving an impetus to change, albeit slower than if the airline had survived.

The State governments are not always competitors in their dealings with the airlines and the tourist bureaux (including QTTC) have funded domestic flights chartered from the airlines. A recent Whitsunday initiative was unusual in having a regional focus. In 1994, Hamilton and Hayman Islands plus Laguna Quays jointly chartered some flights under the banner Whitsundays Cocktail, with a view to securing extra seat capacity at a time when discounted scheduled flights were in short supply. This chartering of flights was an unusual regional co-operative initiative and is highlighted in figure 4.1. Certainly Hamilton Island developer Keith Williams had chartered aircraft during the extreme circumstances of the pilots' dispute, but the recent initiative could only be justified because of the relatively large capacity of Hamilton Island. An equivalent initiative in Fiji would require an unlikely level of co-operation between a multitude of (smaller) properties or an equally unlikely commitment by Government to underwrite such an exercise.

As with domestic packages and as evidenced in table 4.3, the travel agency sector accounts for 85 per cent of outbound air-inclusive holidays. In contrast to the tour operating sector in the UK, the almost complete absence of charter flights has prevented the growth of vertically integrated tour operators and has confined opportunities for integration to the scheduled airlines. Jetset Tours integrated by building up the largest retail travel chain in Australia with 440 outlets. Other wholesalers have evolved from a travel agency base. Thomas Cook, recently purchased by American Express, is the best known example with its tour operating division known as Swingaway Holidays. In 1994, the Commonwealth Bank sold its Travelstrength retail travel arm marking the end of many years of involvement in retail travel by all of Australia's major commercial banks.

Most travel agency chains have resisted becoming involved in tour operations but it is possible that the airlines will allow such retail chains to brand their own tour-based product. Venture Holidays is one of the few genuinely independent tour operators offering a full range of outbound inclusive holidays to destinations including Fiji. Other independent operators tend to specialise in a particular destination, or else a certain type of holiday. Fiji is served by a number of dedicated tour operators, most notably Rosie Tours, Tapa Tours and Fiji Specialist Holidays. The presence of these dedicated operators has been important for Fiji in the absence of any Air Pacific tour operating division. Since the three operators mentioned feature only Fiji, they are unlikely to desert the destination at short notice in the event of tourism problems within Fiji.

Figure 4.1 New holiday concept: Whitsunday cocktail
Because of their larger size, the Whitsundays are better placed to pursue the European model of charter-based holidays.
Source: Barbara Harman, Public Relations and Marketing Manager (courtesy of Hayman Island)

Poon's (1993) view of diagonal integration as a growing trend is manifested in the Australian travel industry. Many players try to achieve synergies between different activities that relate to travel expenditure. Ansett links closely with Diners Club (which it owns), as was demonstrated in table 4.2. Jetset's future has been described as 'aimed more at revenue from communications and money handling than from selling travel in its own right' (Jason 1994: 6). The Managing Director of Thomas Cook has stated that 'the card is the only reason American Express is in the business [of travel]' (Casey quoted in *Traveltrade* 21/9/94). These are examples of Poon's 'diagonal integration'.

Domestic and outbound holiday companies have used different approaches to the classification and grading of resorts. Jetabout and Jetset have used their own resort classification and grading systems for many years to destinations such as Fiji. Jetabout applies its labels of budget, standard, superior, first class and deluxe to all featured properties. The approach is explained by the different resort standards prevailing overseas from within Australia. Classification and grading offered a quality benchmark linked to the credibility of the operator's brand, that is, confidence that such companies would make credible judgements and assessments. Arguably the availability of such labels made it easier for consumers to discriminate between competing resorts.

Until recently, the only significant Australia-wide classification and grading system has been operated by the state motoring organisations and has been directed at independent motorists and not air travellers. Perhaps the perceived familiarity of Australians with the standards pertaining in their own country inhibited any development. As consumer expectations about quality of service have increased, the pressure to deliver minimum and guaranteed service has grown. Sunlover Holidays first applied the Royal Automobile Association (RAA) classifications in the early 1990s. At the time that the interviews were conducted by the author, only Sunlover Holidays used the RAA star rating system and Ansett and Qantas did not. The author sought resort manager opinions about classification and grading because of the obvious contrast evident with overseas resorts such as Fiji where tour operator grading systems are the norm. The author expected that the Qantas/Australian merger might prompt the adoption of the international practice to domestic conditions. In fact Qantas chose to follow the Sunlover example and, for the 1994–5 season, have used the AAA star rating systems in their Queensland Islands and other brochures. At the time of writing Ansett has not implemented a comparable classification and/or grading system. Consequently the responses to questions about classification and grading assumed that any development of this activity would be instigated by Qantas and/or Ansett. Travel agent responses to classification and grading issues are outlined in chapter 5.

Table 4.4 shows the future shape of travel distribution in Australia as forecast by Access Research and provides an insight into prospective relations between agents and airlines. The forecasts depend on the fluctuating fortunes of some fairly volatile commercial organisations but the authors, Access Research, undertake the bulk of the Australian Federation & Travel Agents (AFTA)-commissioned research and are regarded as a reliable organisation. Key trends identified are the growth in agent share by corporate multi-outlet groups from 34 per cent in 1990 to 50 per cent in 2000, a more spectacular drop in the number of independent outlets from 49 per cent to 18 per cent and a doubling of the share of airline-owned outlets from 4 per cent to 8 per cent. The growth of chains points to an improved bargaining strength for the retail sector, but in reality most will have a close affiliation with either Ansett or Qantas. It also points to the growing importance of economies of scale. As travel distribution becomes increasingly polarised, it is imperative for the Whitsundays that Qantas enhances its interest in the region to ensure commitment to selling Whitsunday properties by the Qantas-affiliated network. Welcoming Brampton as an integral part of the Whitsundays might prompt some welcome change. The outlook for Fiji is more challenging still, namely how to ensure that the corporate multi-outlet chains and in particular those affiliated with Ansett, provide sales support and direct their customers towards the Mamanucas. The Mamanucas' positioning as a family-oriented sun, sea and sand destination makes it vulnerable since that market is most prone to concentration in the hands of a few large-scale tour operators. 'Small is beautiful' is a philosophy more comfortably associated with special-interest holidays such as ecotourism, cultural tourism and adventure tourism. None of these activities (nor the companies featuring them) is currently associated with the Mamanucas.

The role of tour operators in marketing the Whitsundays and the Mamanucas does not appear to have faded, despite Poon's (1993) view that tour operators are diminishing in significance. Different reasons do, however, explain the contrasting role of tour operators in the two settings. Tours operating to the Whitsundays has been Ansett dominated, resulting in a lack of variety for consumers with the only substantial non-airline competition offered by Sunlover Holidays. The expansion of resort capacity on the mainland could enhance Qantas' position by making Whitsunday Airport more relevant and convenient as a regional gateway. Some expansion is also evident from competition such as 1-Call. In the Mamanucas Air Pacific has acted in a way which appears to bolster the role of the tour operator, despite Poon's arguments about their relative decline worldwide. By insisting that the cheapest airfares would be available only to consumers booking packages purchased through tour operators, the latter were given significant assistance. The result has probably been more beneficial to Qantas Holidays than to Air Pacific. In defence of the Air Pacific position, one may argue that in view of Qantas' strong position it was desirable to

Table 4.4 Anticipated changes to travel distribution in Australia 1990–2000

Outlet types	1990	%	1995 (est.)	%	2000 (est.)	%
Corporate multi's	1,441	33.7	1,680	43.0	1,850	50.0
Bank multi's	203	4.7	220	5.6	200	5.4
Airline multi's	171	4.0	240	6.1	300	8.1
Flights only multi's	71	1.6	160	4.1	200	5.4
Motoring organisations	95	2.2	100	2.5	100	2.7
Independents (unbranded)	2,073	48.5	1,200	30.7	680	18.4
Independents (UTAG)	216	5.0	300	7.6	370	10.0
Total outlets	4,270	100.0	3,900	100.0	3,700	100.0

Source: Access Research 1992
Note: Assumes 1990 rates of agency openings and closings, and a 20 per cent increase in the current rate of agency mergers and affiliations through franchising

offer incentives (such as exclusive access to the best airfares) to independent tour operators.

Until Air Pacific's recent announcement of its participation in the Qantas Frequent Flyer scheme, the absence of frequent flier associations for Air Pacific was an example of the better positioning of the Whitsundays in terms of diagonal integration than was the case with the Mamanucas. One Mamanuca resort, Mana Island has shown some ingenuity in developing special promotions in conjunction with tour operators, though to date there is little evidence of closer associations with travel agent groups. This type of linkage will be increasingly essential if the resorts are to develop a more targeted approach to their marketing, particularly in view of the greater concentration of travel agency turnover in the hands of multi-outlet groups. Poon contrasted the 'old' tourism of Nassau in the Bahamas with the 'new' tourism of the Family Islands (also in the Bahamas). Such a crude contrast does not apply to the Mamanucas and the Whitsundays. Though the former are smaller scale and less dominated by international operations and the Whitsundays have a competitive strength in a wider variety of market segments, the family is the key source for both destinations.

According to the Access Research forecasts, air packages will account for an increasing proportion of the domestic holiday dollar as cheaper air fares enhance the competitiveness of air travel versus surface transport. The Mamanuca resorts will find domestic Australian destinations more price competitive. The continued loosening of package structures will continue to affect both domestic and international destinations. One danger for the Mamanucas is that Fiji has been seen as a destination suitable for first-time overseas travellers. In most mature generating countries, travellers are favouring long-haul over short-haul travel. As the Spanish island of Majorca has found, traditional short-haul destinations increasingly need to offer a highly distinct tourism experience and not merely the cheapest sun, sea and sand (Morgan 1991). Others can always replicate the latter at a cheaper price, and demand for such products is dropping as a share of total demand in any case. Fiji and its sub-regions such as the Mamanucas need a major marketing impetus to counteract their *passé* image.

This chapter has explained the relationship between the two destination regions, the tour operators, travel agents and airlines. The gradual and continuing shift of Australia's mass tourism 'pleasure periphery' away from Pacific destinations such as the Mamanucas to Asian and domestic destinations has been identified. The Mamanuca resorts are small and do not benefit from economies of scale, whereas the Whitsunday and Asian resorts are larger. Lacking significant air competition and being serviced substantially by Air Pacific with its less sophisticated yield management systems and less developed networks of 'diagonal integration', the Fiji islands have been peripheralised despite their closeness to Australia. Fiji is now competing directly with a developed destination (Australia) which has capital, technological and human resource advantages. The increased fashionability of domestic holidays, not least because of greatly improved standards of food and accommodation, has allowed destinations such as the Whitsundays to take advantage of these resources and to expand their marketing budgets. The Mamanucas are excessively dependent on a declining sector of the industry – wholesalers.

Fiji has little control or influence over industry structure. Air Pacific seeks the respectability of profitability. Whilst apparently welcome, this may not be in Fiji's best interests from a strategic point of view if it means fewer discounted seats. It appears that Ansett and Qantas are the carriers really determining the future of tourism in the 'pleasure periphery' and with Ansett shunning any involvement in the South Pacific, the terms are largely dictated by Qantas.

Overseas-based intermediaries do not have absolute control over demand – clearly consumer preference for particular destinations is important. Fiji's island resorts have been particularly successful in building a strong relationship between staff and guests. Revisitation figures for all Fiji in 1992 were 37 per cent for Australians and 48 per cent for New Zealanders. The lower figure applying to all overseas visitors of 26 per cent was accounted for by long-haul sources such as Europe and North America where most visits are 'one-offs'. The figures are slightly better than those recorded by an emerging Pacific competitor, Vanuatu, which recorded repeat visitation of 21 per cent for holiday travellers (TCSP 1992). Despite the adequate revisitation factor, Fiji authorities should be concerned at the over-dependence on a single source of business – families with young children. Destinations such as Bali also attract families with teenage children. This offers such destinations the prospect of achieving better revisitation figures as families progress through the life-cycle. Given their strong and increasing dependence on overseas-based operators, Fiji resorts must be more proactive in securing repeat business from Australia. As Poon (1993) has said, this is one area where resort operators have the potential to determine their own fate. Over-dependence should not be placed on the growing but more fickle long-haul travellers.

To counteract the risk of losing ground amongst the purchasers of mass market sun, sea and sand holidays and becoming marginalised, the Mamanucas may need a more 'niche marketing' approach. The Mamanucas' traditional source of business is unlikely to grow and the key to success could come from alternative markets such as educational travellers, sporting groups, ecotourists and cultural tourists. In the Whitsundays, changing consumer perceptions have probably exerted a greater influence than actual industry restructuring. Despite Ansett's monopoly of the best airport, the availability of discounted airfares to Cairns and other domestic destinations and the pressure exerted by the several international hotel brands now managing Whitsunday resorts has resulted in a downward pressure on airfares. If mainland resort expansion continues, this has the potential to benefit Qantas and Whitsunday Airport. Clearly the Mamanucas have found it increasingly difficult to compete with the Whitsundays. The group may need to become perceived as a more specialist destination.

Typically of South Pacific Governments, Fiji has regarded the national airline (Air Pacific) as the key to preserving national sovereignty over tourism. Such small national carriers, burdened with a requirement to 'protect the national interest', are increasingly marginalised in an environment where transnational partnerships and cross-shareholdings (as between Qantas and British Airways) are becoming particularly prominent.

One possible scenario that could benefit Fiji would be for the relevant aviation issues to be incorporated within the framework of the CER for the purpose of aviation. Air Pacific could then fly passengers between Australia and New Zealand. The weaker negotiating position of the Fiji government relative to its larger partners makes such an outcome unlikely, however.

The role which Ansett Australia might play is not yet clear. Already operating as a domestic carrier in both Australia and New Zealand, Ansett could apply for traffic rights to Fiji. Thus far at least, Ansett appears to have taken a quite different option with services now offered to Bali, Hong Kong and Japan. The latter two enjoy a high volume of business travel, unlike Fiji and Bali which are predominantly holiday destinations. The entry of Ansett into international routes is a symptom of the growing competitive front which ranges across both domestic and international routes. The focus on Asia is a conscious and deliberate one, suggesting that the South Pacific is a low priority. If Ansett were to fly to the South Pacific, the Ansett Holidays programme would undoubtedly be extended to Fiji since, as previously observed, holiday travel underpins all aviation activity between Australia and Fiji. This would have the effect of breaking the long-established demarcation between domestic and international holidays. The close association between Ansett and Air New Zealand into a grouping with a distinctly Australasian flavour increases the likelihood of such an outcome.

For travel agents, Fiji has enjoyed the advantage of being 'exotic' and importantly, overseas, since this meant that it attracted a minimum of 9 per cent commission. Consumers were also attracted by the high value of the Australian dollar relative to the Fiji dollar (and most other overseas currencies prior to financial deregulation in 1983) and by the availability of plentiful duty-free shopping in Fiji, an option not available to domestic travellers. Nor did Fiji workers demand the type of pay and conditions expected by their Australian counterparts, enabling resort operators to offer more attractively priced accommodation. Finally and quite significantly, domestic travel was simply less prestigious than going overseas. Prior to the late 1980s when the Australian Bicentennial and a number of major international events were staged in Australia, Australians had been coy about buying Australian holidays.

The industry environment has now changed. Fiji still benefits from lower labour costs and from its exotic image. However, Australian destinations are now 'in' with consumers. Following deregulation travel agents benefited from 9 per cent commission levels from Compass Airlines, overrides were paid by Ansett and Qantas/Australian increased their growing lists of preferred travel agents. These facts have disposed travel agents more favourably towards domestic holidays than in the past. Finally, the falling cost of domestic airfares has made Queensland more competitive than previously. If Fiji has been (in Britton's words) a neocolonial, peripheral outpost in the past, it has at least had certain competitive advantages over Australia. It will now be under increasing pressure to compete for customers with Queensland.

5

MARKETING THE PACKAGE

How are the Whitsundays and the Mamanucas promoted in Australia? How similar or different are the methods used and the markets selected? How successful are the destinations at reaching the intended audiences? What about awareness of the actual names Whitsundays and Mamanucas? If there is a difference, does the fact that Fiji is a foreign country and part of the developing world play a part? Do those best positioned to provide detailed product knowledge and advice to intending travellers, namely travel agents, have strong perceptions of individual brands? Are such perceptions stronger than those applying to the island groups as a whole? These are some of the questions that will be addressed in the present chapter.

Possessing a thorough understanding of customers and source markets is critical for any island resort destination. Persuasive communication is also vital, because relatively few island resorts can rely on the stopover or corporate markets to fill their spare capacity – they must rely on the patronage of leisure travellers who have great discretion over their choice of destination. The present chapter uses a marketing strategy framework to assess the relationship between the island resorts – both individually and collectively – and their customers. This detailed analysis should be informative for those with an interest in the marketing of island resorts elsewhere.

In the previous chapter, we examined structural issues impacting upon the various components of the tourism industry. Inevitably the study of structure led us to examine some of the strategies aimed to increase visitation to the two island groups. To compare the marketing effectiveness of the regions amongst Australian consumers, we must study the comparative marketing used to attract them. In comparing the perceived and actual effectiveness of the marketing undertaken by the two destinations one key limitation should be acknowledged. In 1994 the budget of Fiji's peak promotional agency the FVB was F$3.5 million. In 1992–3, the QTTC received over A$19 million from Government, plus A$4.5 million as a 'special' marketing grant and A$3.9 million in industry support. The stark contrast highlights the difficulty of comparing a developing country with a developed one.

Much of the marketing literature in existence prior to the 1980s focused on observations of the manufacturing sector. More recently, a substantive services marketing literature now exists (Lovelock 1984). Services marketing theory has modified some of the key tenets of earlier theory, including the marketing mix. Cowell (1984) extended the traditional 'four P's' of the marketing mix (product, price, promotion and place) to seven, by including the categories of process, physical evidence and people. Morrison (1989) extended the conventional four P's further still to eight by adding the elements of packaging, programming, people and partnership. Middleton (1994) and Holloway and Plant (1992) on the other hand opt for the traditional four. In the present research six P's are deployed, namely the traditional four plus process and packaging.

Destination marketing has been a highly researched field (Ashworth and Goodall 1990; Ashworth 1991). Tourism geographers have related destination choice to consumer decision-making processes through the application of techniques such as the opportunity set method (Goodall, Radburn and Stabler 1988). The study of destination packaging has for the most part been a subset of destination marketing and often prompted by the perceived need by tour operators to select destinations offering the greatest visitor potential and to determine the appropriate capacity. The consumer viewpoint has also prompted a number of studies aimed at assisting consumer decision-making and encouraging truth in destination advertising (National Consumer Council 1988).

In this chapter a straightforward marketing typology is developed to enable comparative analysis of tourism in the two island groups. The typology is presented as a modified version of the diverse models applied to marketing strategy development in various services and tourism marketing texts. The authors of such texts differ in the labels used but the underlying structures that they employ are similar. The technique of drawing selectively from the various elements of marketing strategy and applying it consistently across a variety of examples was used by the present writer in a co-authored book, *Tourism Marketing in Australia* (King and Hyde 1989a). The Hyde and King study included case studies of twenty-six Australian tourism organisations including resorts and relied on secondary sources, supplemented by personal interviews with the heads of tourism corporations. The narrower focus on two specific resort groups taken in the present research has been supplemented with a wider range of surveys and interviews with consumers, travel agents and destination marketing authorities. Useful data on comparative holiday costs has also been gathered by examining promotional materials, particularly brochures. In the present chapter, the travel agent survey findings are the primary source. Since respondents constituted a representative sample of the total population (travel agents in Melbourne and Sydney) more confident conclusions could be drawn. The typology involves a four stage process and incorporates a number of subsidiary elements as outlined in table 5.1.

ASSESSING THE ATTRIBUTES OF ISLAND RESORTS

The first of the four stages is the determination of a SWOT analysis. The assessment of strengths, weaknesses, opportunities and threats, or SWOT, is an essential preliminary to the development of a marketing strategy or plan (McDonald 1984). It functions as a bridge between the external environment within which an organisation operates (usually depicted as opportunities and threats) and internal resources and considerations (depicted as strengths and weaknesses) and is placed early in the strategic process. SWOT analysis involves an assessment of the present circumstances which confront a given organisation and its future prospects.

As far as the author is aware, no SWOT analysis has been prepared for the Mamanucas specifically. This explains the use of the Plan covering all of Fiji as the point of reference. The Fiji Marketing Plan (FVB 1993a) includes a separate SWOT analysis for each geographical target market of which Australia is one, unlike the WVB Business Plan (WVB 1993) which relies on the more conventional overall SWOT analysis.

A number of differences are evident between the ways that the two documents treat their respective SWOT analyses. The Whitsunday listing is fairly detailed, whereas the Fiji listings are brief (sixteen versus four strengths, twenty versus two weaknesses, seventeen versus two opportunities and fourteen versus three threats). Whilst the significance of individual numbers should not be over-emphasised, the FVB attention to this aspect of strategic marketing is clearly cursory relative to the Whitsundays. This observation was reinforced by the then Head of Marketing at the Tourism Council of the South Pacific (TCSP) (S. Lolohea pers.comm.), who commented that the FVB was essentially tactical and lacking concrete strategic objectives.

Table 5.1 Marketing typology for comparing the Whitsundays and the Mamanucas

SWOT analysis	Threats and opportunities – external environment Strengths and weaknesses – internal environment
Positioning	Determining the relationship between various brands
Target marketing	Determining a variety of significant market segments
Marketing mix	Product Price – perceived and actual Promotion Place / distribution Packaging Process – technology

The main focus of the Fiji SWOT analysis is the need for increased promotional expenditure and a greater emphasis on clearly identified target markets, notably the 'expanding eco/cultural/heritage tourism and adventure segments' (FVB 1993a: 67). The Whitsundays document places more emphasis on the relationship between marketing and other issues such as investment, product development and planning. The WVB document, acknowledging a previous lack of co-operation, shows a commitment to regional identity and co-operation not evident in the Fiji document.

Tables 5.2 and 5.3 record the responses of travel agents to survey questions about the strengths and weaknesses that they perceive as applying to the two regions. As indicated in table 5.2, travel agents perceived the key strength of the Whitsundays as its accessibility with location and distance collectively attracting 55 per cent of respondents. The other significant strengths were activities and beaches. The Mamanucas elicited contrasting responses. Culture was cited as the main strength. Other significant categories were beaches, price and activities.

In the Whitsundays, the dominant weakness was price (71 per cent of respondents) with few citing it as a strength. The perception of high price is clearly an issue that the Whitsunday tourism authorities should address. In the Mamanucas price was also seen as the most significant weakness

Table 5.2 The Whitsundays and Mamanucas compared: key strengths

| Feature | Whitsundays | | | Mamanucas | | |
	Frequency	Percentage	Feature	Frequency	Percentage
Location	80	40.6	Culture	76	39.6
Beaches	37	18.6	Beaches	33	17.2
Activities	35	17.1	Price	30	15.6
Distance	28	14.1	Activities	28	14.6
Others	18	9.6	Distance	19	9.9

Source: Author's survey of travel agents, December 1993
Note: n = 200

Table 5.3 The Whitsundays and Mamanucas compared: key weaknesses

| Feature | Whitsundays | | | Mamanucas | | |
	Frequency	Percentage	Feature	Frequency	Percentage
Price	116	70.7	Price	47	31.3
Beaches	16	9.8	Distance	39	26.0
Activities	12	7.3	Beaches	21	14.0
Culture	6	3.7	Activities	14	9.3
Distance	5	3.0	Food	14	9.3
Others	9	5.5	Others	15	10.1

Source: Author's survey of travel agents, December 1993
Note: n = 200

though by less than a third of all respondents. Unlike in the Whitsundays where it was seen as a strength, 26 per cent mentioned distance as a weakness for the Mamanucas. This perception reinforces the view that the Mamanucas are less accessible than the Whitsundays. Activities and food were mentioned as weaknesses by a small but significant number of respondents. The Mamanucas are positioned slightly less favourably than the Whitsundays and the two issues should be monitored closely by the Mamanuca authorities. Activities were regarded a strength in the Whitsundays and whilst food was less positively regarded, only one respondent mentioned it as a negative. Activities and food are significant elements in consumer perceptions of overall 'quality'. Table 5.3 outlines the weaknesses of the two areas as perceived by travel agent respondents.

AWARENESS AND VISITATION

Since most holiday bookings to the two island destinations are made through travel agents, it is appropriate that we develop an understanding of travel agents' attitudes. The present section focuses on agent familiarity and visitation. In most consumer behaviour models, awareness is only the first element in a multi-stage process. Two of the best known models are Awareness–Interest–Desire–Action (AIDA) and Attention–Comprehension–Attitudes–Intention–Purchase (Howard and Sheth 1969). Awareness and previous visitation are determinants of whether travel agents will recommend a particular destination to potential customers. Table 5.4 outlines agent responses to the question: 'Which island resorts in Queensland and Fiji come first to mind?' The smaller and usually less well-known resorts recorded most non-responses. Non-respondents to particular questions were assumed to be unaware of the specific resorts affected.

Interviewers were asked to record agent responses verbatim to the question 'which island resorts do you think of?' Total responses are included under the heading 'unprompted awareness'. Some resorts outside the Whitsundays and the Mamanucas are listed in table 5.4 as mentioned by respondents. Some under-reporting could have occurred, since only islands in the latter two groups were referred to directly in the questionnaire.

The Qantas Australian resorts, Dunk (eighty-two mentions) and Great Keppell (seventy-two mentions) received the most mentions. Qantas also operates the two other highest profile Queensland properties in the list, namely Lizard and Bedarra islands, suggesting that Qantas' main Queensland island focus is outside the Whitsundays. The fact that these islands are backed by the marketing muscle of Qantas makes them a significant source of competition. Mamanuca resorts enjoy a near monopoly of unprompted agent awareness of island properties in Fiji. Whilst the small, deluxe Turtle Island in the Yasawas adjacent to the Mamanucas recorded significant

Table 5.4 Queensland and Fiji island resorts: unprompted travel agency awareness

Island resort	Mentions	Island resort	Mentions
Dunk	82	Heron	22
Great Keppell	72	The Fijian	9
Lizard	29	Orpheus	7
Turtle	26	Vatulele	7
Bedarra	25	Fraser	6

Source: Survey of travel agents, December 1993
Note: n = 200

recognition, no other Fiji property attracted comments from ten or more respondents.

The insignificant recognition factor of the larger Fiji properties outside the Mamanucas may have arisen from the fact that they are not strictly island resorts (the Regent and Sheraton hotels on Denarau Island, Sonaisali Resort and the Fijian resort on Yanuca Island are all connected to Viti Levu by causeway). Overall, Queensland islands outside the Whitsundays enjoy a much higher profile and awareness as island resorts amongst travel agents than do Fiji resorts outside the Mamanucas. The Mamanucas and the nearby Yasawas are the only island-style properties which recorded significant agency awareness. The relatively high concentration of the most familiar island destinations could lend itself to a strong campaign by the Mamanucas to emphasise their special allure as a complete island destination, easily accessible from the country's major gateway. Respondents were subsequently asked about each individual resort name except for those previously mentioned and their responses are listed 'prompted awareness'. Some overstatement of recognition could have occurred. Non-awareness is recorded as 'unaware'. Agents were then asked which resorts they had visited.

As previously stated, Hamilton Island has greater room capacity than all the Mamanucas combined, so not surprisingly it was the most visited island as indicated in table 5.5. The occupation of second and third place by Plantation (110 rooms) and Castaway (sixty-six bures) respectively is more surprising. These resorts must have emphasised travel agent familiarisation trips (most of the visitation cited is assumed to have been at least partly educational). Just over one-fifth of respondents had visited Daydream, Mana and South Molle islands with Treasure Island recording just under one-fifth. It is notable that all of the medium to large Mamanuca resorts (say those with in excess of fifty units) have hosted substantial numbers of travel agents. In view of the much larger average and overall size of the Whitsunday resorts, it is clear that the latter groups have neglected travel agent educationals in relative terms. The low figure for Hayman Island probably reflects the latter-day conversion of the resort to sales through

113

travel agents. Hook Island and Palm Bay Hideaway recorded the lowest visitation, the lowest unprompted awareness and the highest level of ignorance in the Whitsundays. Low visitation to Club Med may be explained by the newness of the resort. The name Radisson Long Island was unfamiliar to agents and the name of that resort has changed once more. The issue of agent awareness needs urgent attention by new operators, the Club Crocodile group. Given that the Club Crocodile name is best known for its established Airlie Beach property, communicating the differences and similarities between the island and mainland properties will be a significant challenge. The fact that cross-ownership has arisen must be viewed as a bonus for overall integration between the on- and off-shore areas.

Though a cost benefit analysis of travel agent visitation and awareness relative to sales is not possible here, it should be observed that the Whitsundays appear to have neglected travel agents relative to their room capacity. This neglect is symptomatic of the historically lower propensity of travel agents to sell domestic holidays. The greater enthusiasm now being shown by Hayman and others suggests that the Mamanucas may

Table 5.5 Agent awareness and previous visitation (in percentages)

Resort	Unprompted awareness	Prompted awareness	Have visited	Unaware
Hamilton	65.5	34.5	35.5	–
Plantation	52.5	43.5	33.0	4.0
Hayman	47.5	51.5	15.5	1.0
Daydream	41.5	57.5	21.5	1.0
Castaway	41.0	55.0	25.0	4.0
Mana	33.0	58.0	21.5	8.0
South Molle	32.5	63.0	21.5	4.5
Brampton	30.0	66.0	11.5	4.0
Treasure	26.5	68.5	18.5	5.0
Club Med	18.0	77.5	10.5	4.5
Beachcomber	17.0	81.0	17.5	2.0
Matamanoa	15.0	65.7	12.0	19.5
Radisson Long Island	12.5	79.0	7.5	8.5
Naitasi	9.0	65.5	13.5	25.5
Sheraton Vomo	8.0	56.0	3.0	36.0
Navini	6.5	67.0	8.0	26.5
Tavarua	5.5	57.5	4.5	37.0
Tokoriki	5.0	54.5	5.5	40.5
Musket Cove	4.5	81.0	9.0	14.5
Hook	3.0	71.0	6.5	26.0
Palm Bay	0.5	49.0	3.5	50.5

Source: Survey of travel agents, December 1993
Notes: Totals exceed 100 because of overlapping categories visited and unprompted awareness
n = 200

face intensified competition from the Whitsundays in this sphere. Given the over-dependence of the Mamanucas on tour operators, it is essential that their hard work with travel agents is sustained. A close relationship with travel agents is a key element in maintaining a relationship with customers in source markets.

The smallest resorts registered the lowest agent awareness – Tavarua because of its specialist market (surfers), Vomo its newness and Navini its tiny scale. The poor awareness levels for Palm Bay, Hook, Naitasi, Tokoriki and Matamanoa all merit closer scrutiny by the resort operators. The management at Hook Island was uninterested in attracting Sydney- and Melbourne-based travel agents. The new operators of Hook Island, the Jewel Group, should consider a closer association with travel agents, given the shortage of 'informal-style' accommodation in the Whitsundays. Some upgrading of existing facilities would probably be needed.

POSITIONING

Market positioning may be defined as the process of establishing and maintaining a distinctive place in the market for an organisation and/or its individual product offerings. It draws upon SWOT analysis in incorporating an assessment of internal resources and of competitors and adds a marketing research dimension through relating these two elements to an understanding of consumer perceptions and preferences. In the present study, competition for the various resorts is assumed to come from other island resorts in the Mamanucas and the Whitsundays. The pilot travel agent survey had suggested a minimal agent awareness of the Mamanucas name but a better awareness of individual resorts, vindicating the decision to examine certain individual resorts in more detail. The analysis of correspondence (ANACOR) technique of correspondence analysis was used to produce a graphical and comparative depiction of the six resorts: Hamilton, Daydream and Club Med Lindeman islands in the Whitsundays, Plantation, Castaway and Mana islands in the Mamanucas.

Figure 5.1 highlights the qualities most strongly associated with the Mamanucas: intangibles such as 'atmosphere', notably 'romantic', 'exotic', 'different', 'relaxing' and 'friendly'. Tangible associations were weaker with facilities regarded as 'basic'. Mana is located close to the label 'basic facilities' and closest to the label 'exotic' and (with Castaway) 'romantic'. Mana clearly needs to attend to its facilities as a matter of urgency. Daydream and Club Med recorded strong associations for facilities and service. The current installation of better quality accommodation at Mana (deluxe bures) will need to be accompanied by an overall upgrading in facilities and service, if the prospective higher-spend guests are to be satisfied. The label 'value packages' is equidistant between the Mamanuca resorts and their Whitsunday equivalents (Daydream and Club Med). Agents perceive that

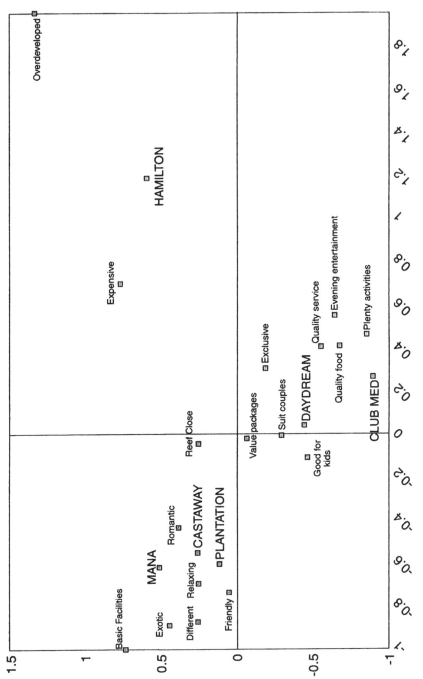

Figure 5.1 Brand map: correspondence analysis

neither destination has a clear advantage. Hamilton Island is perceived as 'expensive on the island'.

The recently rebuilt Daydream and Club Med resorts offer highly capitalised and up-to-date features and fixtures. Both are well positioned as destinations suited to a variety of market segments ('good for kids, suit couples') that offer a range of distractions ('evening entertainment, plenty activities') and quality service ('quality service, good food'). Hamilton is dominated by a perception of overdevelopment and overpricing. Of the six, it is furthest from the congenial atmosphere of the Mamanucas ('exotic, different, relaxing and friendly'). The results indicate poor market positioning for Hamilton Island (see also figure 5.2 for an illustration).

The correspondence analysis exercise points to features that the properties could emphasise to best harness agent preferences. Agents perceive the individual islands as sharing similar characteristics with other resorts in the same group and demonstrating more contrasts with members of the other group. Hamilton Island is the exception in being regarded as very different from other Whitsunday resorts. Of most concern for Hamilton Island are its perceived deficiencies on issues which are fundamental to the popular concept of the island resort experience.

Figure 5.2 Hamilton Towers
The towerblock as 'spectacle'. Hamilton Island Resort 'towers' over the competition, but its brazenness is an irritant for many visitors and for travel agents.
Source: Nicole Witzgall, Executive Assistant, Hamilton Island Resort (courtesy of Hamilton Island Resort)

TARGET MARKETING AND MARKET SEGMENTATION

Market segmentation is the process of dividing a population into several different groups, each displaying similar product interests and relative internal homogeneity (Kotler 1991). Segmentation has become a prerequisite in modern marketing because consumers are showing increasing resistance to being treated as undifferentiated elements of a 'mass market' and are seeking products and experiences more precisely tailored to their perceived needs (Poon 1993). Target marketing entails the selection of those segments considered best suited to a particular category of product.

Segment descriptions were arrived at which best highlighted the similarities and contrasts between the two destinations. Question 3 in the travel agent survey asked respondents to assess the suitability of the two island groups for a variety of predetermined market segments. The earlier evaluation of strengths and weaknesses was extended to particular groups of consumers, as perceived by travel agent respondents focusing on island groups and not on individual resorts. Respondents had already been made aware of the make-up of the two groups which was vital in the case of the Mamanucas, since respondents showed lower awareness of the group name. The ANACOR technique was used to analyse the relevant data; these are recorded in figure 5.3.

The Mamanucas offer greatest appeal to those interested in an alternative culture, to honeymooners, to those in search of the 'classic sub-tropical holiday experience', to 'stressed' executive couples and to sociable middle-aged couples. From this positioning, we may conclude that cultural attributes are a clear (and obvious) advantage and that is the aspect that most conspicuously differentiates the Mamanucas from the Whitsundays. The Mamanucas appear to be maintaining their traditional appeal for honeymooners though Queensland now produces special honeymooner brochures, enabling the Whitsundays to promote through specialist distribution channels such as wedding magazines. A more insistent regional and Queensland emphasis on honeymooners could threaten one of the Mamanucas' most established sources of business. It is notable that Daydream and Brampton islands have been increasing their family emphasis, relative to honeymooners. A perception is evident that an island does not need to confine itself to a single target market, but can use specialist channels to attract a variety of segments.

The agency perception that Fiji better fits the 'classic sub-tropical island resort' image suggests that domestic island resort destinations have not shared the imagery of the Pacific Islands. Though the Mamanucas appear well positioned with 'stressed executive couples', the 'basic facilities' image shown in the previous correspondence chart could lead to dissatisfaction amongst this group. It would also constrain the ability of the Mamanucas to tap the conference and meeting organisers who typically demand high

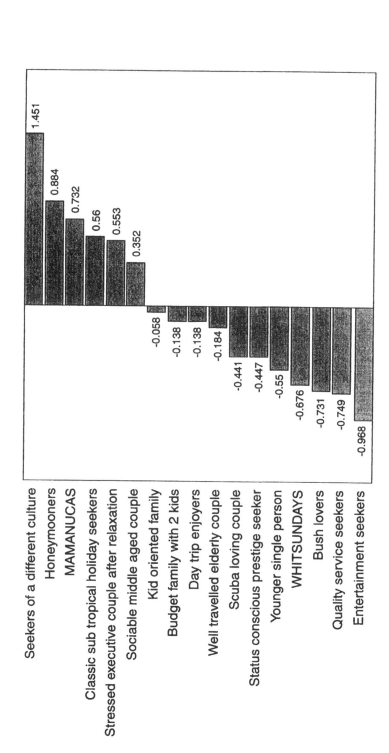

Seekers of a different culture — 1.451
Honeymooners — 0.884
MAMANUCAS — 0.732
Classic sub tropical holiday seekers — 0.56
Stressed executive couple after relaxation — 0.553
Sociable middle aged couple — 0.352
Kid oriented family — -0.058
Budget family with 2 kids — -0.138
Day trip enjoyers — -0.138
Well travelled elderly couple — -0.184
Scuba loving couple — -0.441
Status conscious prestige seeker — -0.447
Younger single person — -0.55
WHITSUNDAYS — -0.676
Bush lovers — -0.731
Quality service seekers — -0.749
Entertainment seekers — -0.968

Figure 5.3 Correspondence analysis: market segmentation

standards of service. The travel agent perception that the Mamanucas are well positioned to cater to short-break travellers could be a positive for the 'stressed executive couples' market, though lack of interest by the FVB may prove an obstacle. The Mamanucas were seen as suitable for 'sociable middle-aged couples', reinforcing the focus group finding that friendliness was a key strength for Fiji.

The Whitsundays were suited to those in search of the types of activities and facilities reported as being well catered to in the previous correspondence chart, namely 'entertainment seekers' and 'quality service seekers'. The links with the national parks are already appreciated by travel agents who regard the Whitsundays as suited to 'those who like getting out into the bush', though the Australian association of the term may have reduced the number of Mamanuca responses.

The heavy marketing by Hamilton and Hayman to 'status conscious' consumers in the early days after their respective redevelopments may help to explain the strong association achieved by the Whitsundays with that group. Executives not preoccupied with status and prestige and in pursuit of total relaxation should be a primary target for the Whitsundays. Perhaps the perception that Hamilton is not relaxing has carried over to perceptions of the Whitsundays as a whole. Why, though, were the Mamanucas seen as better suited to 'stressed' executive couples? The Whitsundays' relative failure (in perceptual terms) with this group is an indication that regional marketing may need to modify some earlier Hamilton and Hayman induced imagery.

The Whitsundays are well placed in the dive market, perhaps partly prompted by the devotion of ten pages in the QTTC's *Dive Queensland* brochure. The strong performance by the Whitsundays with divers is surprising in that most diving involves a lengthy transfer of up to two hours from the relevant resort to the Great Barrier Reef. In contrast, most of the Mamanuca islands offer reef diving close to shore. The greater success of the Whitsundays in the dive market indicates the potential of highly targeted market segmentation reinforced by specific promotional materials. Fiji clearly needs a stronger marketing campaign targeted at divers. The Whitsundays also have an edge with families (including the budget conscious), younger singles and well-travelled elderly couples. The strong showing of the Whitsundays with families should be of some concern to the Mamanucas, especially in light the recent expansion of state-of-the-art childcare facilities in the Whitsundays. This issue is discussed in detail in chapter 7.

Focus group respondents had cited 'being trapped' as a negative connotation of island resorts. Travel agent respondents suggested a different conclusion with only 12 per cent of respondents viewing both groups as unsuitable for holidaymakers who 'enjoy a variety of day trips'. Agents appear to regard island resorts as offering considerable mobility.

Generally, agents regarded the Mamanucas as catering less effectively to a range of market segments. It recorded a positive profile as a 'honeymooner' and 'cultural' destination and as a place for a 'sociable middle-aged couple'. Even here though, the evidence is not wholly positive. The *sociable* middle aged couple will be satisfied but, according to travel agents at least, the *well-travelled* middle aged couple would be better served by the Whitsundays. This reinforces a criticism made that Fiji is a 'first-time' overseas destination. The result is reassuring for the Whitsundays, since it suggests that destination is preferred by discerning mature-aged customers. The Mamanucas are particularly uncompetitive, according to travel agents, amongst 'younger singles', 'entertainment seekers' and those concerned with 'high quality service'.

As evidenced by resort manager comments and by the type of promotional material used, the Mamanucas' target marketing is oriented towards families and couples. In some resorts the policy is not targeted at all, but is simply regarded as a 'sensible' or cautious approach, since families and couples are seen as the mainstay of mainstream resorts throughout the world. Dick Smith of Musket Cove resort (pers.comm.) described his key sources of business as 'anybody – we get a bit of everything'. Though Smith has been a key contributor on those occasions when the Mamanucas have conducted co-operative, highly targeted campaigns, his attitude to overall targeting is symptomatic of a lack of focus throughout the group.

The most explicitly family-oriented resorts are Plantation, Castaway, Tokoriki, Mana and Treasure islands. Few resorts, however, limit their marketing to families and no family-oriented island actively discourages couples without children. Sheraton Vomo with its focus on 'exclusivity' is an exception. The overall attitude of management was summed up in the parting words 'Our forward bookings are excellent – very few with children, fortunately' (N. Palmer pers.comm.). Children are not actively barred from the resort but the image of 'exclusivity' communicates a strong, if subtle, message.

Most resort managers stated that families and couples can 'co-exist' in relative harmony. There is also a business logic for the 'co-existence' approach. Travel agents acknowledged the established reputation of the Mamanucas as suitable for honeymooners. In the case of Elegant Tropical Island Resorts (Fiji) Ltd, leaseholders of Matamanoa and Tokoriki islands, management have targeted the former resort at couples and particularly at honeymooners (children under 13 are excluded) and the latter at families (H. Steinocker, pers. comm.). The two resorts were deliberately designed this way with children's play facilities constructed at Tokoriki but not at Matamanoa. Navini is currently targeting honeymooners to diversify its base, and one of the eight bures has been redesigned to appeal to that group. The social balance between families and honeymooners is a sensitive one, though apparently managed with relative success to date in the

Mamanucas. The issue of relations between various groups on the islands is further developed in chapter 7.

Two other Mamanuca resorts are targeted away from families. Tavarua is geared to 'surfers', though families are not actually discouraged and the tariff policy does accommodate children at reasonable rates. Beachcomber is targeted at the 'young at heart' (Kalo Tuai and Akanisa Dreunimisimisi pers.comms.), with its emphasis on lively activities, bands and parties. Again it does not discourage families, but is more focused on couples and singles with its dormitory accommodation (see figure 5.4).

Except for Tavarua, there is comparatively little deliberate target marketing by individual resorts. Where special interest groups are attracted, it is almost accidental. Musket Cove gets 'an active clientele because we market our activities' (D. Smith pers.comm.). The process of market segmentation is seen by the resort managers as mainly about developing strategies for targeting particular source countries. Certain resorts emphasise the European market (for example, Matamanoa), others the Japanese (Mana), some the New Zealanders (Naitasi), and Tavarua the Americans. With the exception of Tavarua (where the US is the major source of business) and Naitasi (where it is New Zealand), all resort managers stated that historically Australia has been their largest single source of business. The declining share of business accounted for by Australians is taken up in chapter 7. The Mamanucas could now benefit from collective promotion highlighting the different features of each resort and the types of client most likely to be attracted to each.

Daydream previously targeted both couples and families. According to Vincent (pers.comm.), 'next year [1994] we will be marketing ourselves more as a family resort', reasoning that whereas it is straightforward for unencumbered couples to travel overseas, domestic travel may be the only travel option for families. Mahony (pers.comm.) described honeymooners as accounting for 40 per cent of total business at Brampton with over-45 'empty nesters' also important. Families were described as previously neglected but now more emphasised through the transformation of the resort's image away from 'romantic' (with its particular appeal to honeymooners) to 'discovering oneself' (which can appeal to any market segment, including families).

Hamilton Island Resort has always attracted a diverse clientele including families, but its image of 'exclusivity', yachts and media personalities helped to create the image of an island out of the reach of 'average families'. The island now offers child care facilities and promotes an image that the island is suited to 'average families and average people' (S. Cogar pers.comm.). The findings of the present study suggest that the message may have failed to reach travel agents and consumers. Club Med Lindeman Island targets both couples and families. According to the Chef de Village (B. Giampaolo pers.comm.), 'families are particularly important for us

LIVE LIFE TO THE FULL...

Figure 5.4 Live life to the full at Beachcomber Island
Though you can 'Get Wrecked', as you could at the old Great Keppel Island in Queensland, this is the youth market Fiji style. Partying, Fiji-style, is inclusive of all age groups (the 'young-at-heart').
Source: Akanisa Dreunimisimisi, General Manager, Beachcomber Island Resorts and Cruisers (courtesy of Beachcomber Island Resorts and Cruises)

during school holidays. It's a vital market for us'. Club Med places particular emphasis on organised activities. The smaller resorts of Palm Bay Hideaway and Hook Island Wilderness Lodge do not offer special facilities for children, but both attract families in search of a destination with fewer structured activities.

THE MARKETING MIX

In most strategic marketing models, the application of the marketing mix follows after the selection of target markets. The marketing mix is a method of classifying the marketing activities directly under the control of management ('controllable variables'), as opposed to 'uncontrollable variables' such as technological change, shifts in industry structure and the state of the international economy. Six elements of the marketing mix – product, price, promotion, place, process and packaging – will now be considered in turn. Question 5 in the travel agent survey sought respondent views towards a variety of propositions about the marketing mix using a five-point Lickert scale. The results are noted in table 5.6.

Despite making progress towards an acknowledgement of the multi-faceted nature of the intangible service 'experience', the approach of marketing theory to *product* is limiting when we attempt to examine resorts in a holistic way. Island resorts may be pure products sold to and purchased by consumers but there are many social, cultural and environmental dimensions which remain inadequately accounted for by the 'recipe' approach of product as a single element of the marketing mix. Even apparently wide-ranging definitions such as Baker's view that 'product is everything the consumer gets for his or her money' (Baker 1985) fail to answer how one accounts for intangible dimensions such as interactions with local residents and environmental quality. In later parts of the present study, aspects of product are dealt with in the chapters on image, social dimensions and environmental dimensions. In this section particular emphasis is placed on perceptions of product 'quality', recognising that quality is difficult to quantify since tourism products are 'experiences' rather than readily measurable goods.

The proposition in table 5.6 receiving strongest affirmation from travel agents was the extent to which the two groups matched the expectations of clients sent there, namely 91 per cent for the Whitsundays and 86 per cent for the Mamanucas. Agents appear broadly satisfied with the products offered by the two groups, though slightly more so in the case of the Whitsundays. The poorer Mamanucas result could relate to issues of service quality. In the various open questions, a number of travel agents commented that the facilities, services and activities provided in the Mamanucas were of lesser quality than those in the Whitsundays. Whilst the slower staff service in Fiji was compensated for by the relaxed island

Table 5.6 Travel agent agreement/disagreement with propositions about the Whitsundays and Mamanucas

Proposition	Agree a little / a lot	Disagree a little / a lot	Neither agree nor disagree	Non response / can't say
Believe the Whitsundays have matched the expectations of clients who they have sent	90.5%	5.0%	3.0%	2.0%
The use of a resort grading system would help the Whitsundays	72.8%	24.2%	2.5%	1.0%
Would rather recommend the Whitsundays for a short relaxing holiday	79.5%	16.0%	4.0%	0.5%
Information is easier to obtain on the Whitsundays	39.0%	49.5%	7.0%	4.5%
The Whitsundays offer better incentives to travel agencies	44.5%	51.0%	3.0%	1.5%
Recommend the Whitsundays in preference to the Mamanucas	32.0%	44.5%	21.0%	2.5%
Believe the Mamanucas have matched the expectations of clients who they have sent	85.5%	6.5%	3.0%	5.5%
The Mamanucas are harder to sell because of consumer ignorance of the destination	45.5%	53.0%	1.0%	0.5%
Believe the Mamanucas to be cheaper	36.4%	53.0%	7.6%	4.5%
Higher commission make the Mamanucas more attractive to sell	27.0%	67.5%	5.0%	0.5%
Believe that the Mamanucas offer a wider range of good value packages	32.0%	60.5%	6.0%	1.5%
Clients leave the choice of tour operator to the travel agent when booking these destinations	83.5%	13.0%	3.0%	0.5%

Source: Travel agent survey, December 1993
Note: n = 200

environment and friendly disposition of the Fijian employees, the poorer quality was also compensated by the 'bure on the beach' settings. The typical smaller scale and more secluded nature of the developments were also seen as positives.

There was a general suspicion about standards in developing countries. The comment of one agent who specialised in the older-age market segment highlighted concerns about the poorer quality of medical support likely to be available in a developing country. A perceived lack of safety practised on inter-island launches was commented upon about the Mamanucas but not the Whitsundays. So far, the Mamanucas seem to have thrived on the allowance being made by travel agents for poorer service and facilities. To some extent such allowances will always be made, but the substantial upgrading of facilities in the Whitsundays (described in chapters 3 and 8) is a major challenge.

A significant number of respondents stated that Fiji needs to reassure clients about standards and that an accommodation grading system would be the best way to do this. No such desire was expressed by respondents about the Whitsundays. Whilst Jetset and Jetabout already apply a classification and grading system to Fiji resorts, agents would like to see the practice extended to the remaining operators and to the destination authorities. The increased use of grading systems for domestic resorts should prompt Fiji and the Mamanucas to develop a strategy for providing consumers with reassurance about the quality and level of service and facilities offered in the resorts.

Price is the only element of the marketing mix which makes a contribution to revenue since the other P's constitute overheads or expenses for the business. Correct pricing of products can ensure the generation of adequate revenue to cover expenses and make a profit. In resort islands, price consists of a variety of elements, namely transport (by air and including transfers), accommodation at the resort (which may or may not include food and beverage costs) and other expenses such as souvenirs and activities, though many activities are included in the up-front holiday cost. Conventional tourism thinking suggests that short-haul sun, sea and sand tourism is highly price sensitive and that alternative destinations are highly substitutable. Is price the key element to distinguish the relative competitiveness of the Mamanucas and the Whitsundays?

Actual costs are compared with costs as perceived by travel agents and consumers. A comparison of the three principal cost elements of accommodation, airfares and food revealed that the Whitsundays are on average 18 per cent cheaper. Whilst a margin for error of up to 20 per cent could have resulted from the many assumptions implicit in the assessment, it is clear that the consumer perception that the Mamanucas are 'much cheaper' is inaccurate. Travel agents have a more accurate view but may also be guilty of overestimating the relative cost of a Whitsunday holiday.

For consumers, it is likely that perceived quality, rather than absolute price will emerge as the determining issue, though attractive lead-in prices will continue to be a potent means of generating additional business. As Poon (1993) has said, consumers are looking for excellent quality but at a middle price. This consumer expectation will continue to be both a marketing and a product delivery challenge for both destinations. Island promotions typically emphasise price and 'value for money', as evidenced in figure 5.5.

Promotion is 'a combination of advertising, sales promotion, merchandising, personal selling and public relations/publicity' (Morrison 1989: 287). Promotions are deployed because 'knowledge about the product has to be communicated to the potential customers, either through word of mouth, advertising or some form of display' (Holloway and Plant 1992: 96).

One agent characterised the promotion of Fiji as 'terrible – people don't know enough about it'. No aspect of Fiji promotions emerged unscathed – respondents want more educationals, more television and radio advertising, more visits to travel agents by sales representatives, more image building and a fresher image generally, more detailed information about the region and about individual resorts! In contrast a typical comment about the Whitsundays was that 'the QTTC have got it all worked out, especially as the Whitsundays are already well known'. In contrast, only one major deficiency was evident for the Whitsundays – the lack of educationals. Many respondents wanted more educationals from both island groups, perhaps a predictable response from travel agents who regard such visits as their major perk.

Information was perceived as lacking in current Fiji promotions. Referring to the Mamanucas, one Sydney respondent stated that travel agents need to know 'what's available other than beaches – trekking, diving and boating activities'. Another said that 'it's boring as there is no sightseeing – it's only sun and sand'. The serious threat posed by the perceived absence of information for agents is compounded by consumer ignorance about the Mamanucas. There is a perception that because the Mamanucas are secluded, there must be little to do. Some described it as a destination for honeymooners, namely too quiet to satisfy other consumer groups. Others again see it as a family destination, with insufficient attractions for singles or for couples. This indicates a fuzziness of image. If adequate activities are available, they are certainly not well known. Apart from honeymooners, respondents did not cite any particular segment as being well serviced by the Mamanucas. This suggests that what little niche marketing has been undertaken to date has not succeeded in shifting the overall perceptions of the destination. The easy accessibility of the

Figure 5.5 South Molle Island gives you 'the lot'
Highlighting too many resort attributes may detract from the overall idea of an island experience.
This type of promotion is also susceptible to comparison with cheaper mainland alternatives. At
$238 per couple per day, it is costly for low income earners.
Source: John Grocock, General Manager, South Molle Island (courtesy of South Molle Island
Resort)

Mamanucas to mainland activities such as trekking and rafting as well as to the main airport has not been communicated effectively.

Many respondents regarded Fiji as over-dependent on wholesalers, with insufficient reinforcement by the FVB and the islands themselves. The 'FVB', wrote one Melbourne agent, 'could have general information on the islands not linked to any wholesaler'. The practice of tour operator brochures to emphasise resort facilities but not activities and general destination information, was perceived as a deficiency. A Melbourne respondent requested 'tourism information, not just details on accommodation prices and amenities'. According to one Melbourne agent, the FVB 'needs to raise the profile and each [Mamanuca] island needs to be distinguished'. The confined use of grading systems by only the largest operators was regarded as insufficient by respondents, who wanted a better guide to quality and requested that the Fiji authorities introduce a grading system to facilitate comparisons with other destinations. Too much emphasis was placed on hotel information with less on activities and general conditions. 'Let people know more about what they can do and the different tours that are available', said a Sydney agent. Whilst not acquitting the FVB of responsibility, the information gap highlights the dilemma of tourism in resource-short developing countries. Do such countries rely on foreign companies to do their marketing or spend extra government funds that could be diverted to key areas such as health, education and transport?

One Sydney respondent said that Fiji 'needs a higher profile'. Other comments included 'it needs a higher profile and more aggressive marketing', 'people don't know enough about it', and 'the FVB should advertise direct to the public'. Repeatedly, Fiji's marketing was described as ineffective and inadequate relative to Queensland's. A Melbourne respondent stated that 'the FVB should advertise their specials more on TV like Qantas and Ansett are doing'. A Sydney respondent said that the Mamanucas 'need to have a serious look at the Whitsundays and make sure they are cost effective and cost competitive'. Another from Melbourne said that the FVB 'should produce more brochures similar to those produced by the QTTC' and highlighted the need to raise the profile of sub-regions such as the Mamanucas. Comparisons were also made with Bali (already mentioned) and with another Pacific destination, Tahiti. A Melbourne respondent stated that 'unlike in Fiji, we have a book on Tahiti which tells us everything about it'. The comparison with French Polynesia is instructive. That destination emphasises the identity of the various island groups such as the Marquesas, which make up the total country under the slogan 'Tahiti and her Islands'. Tahiti's campaign followed a major slump in tourism traffic to the country. Fiji requires an equally drastic approach.

Queensland's promotional activity was generally viewed as satisfactory. Most criticisms focused on price, an issue largely outside the jurisdiction of the Queensland government authorities. The Whitsundays also appeared to be benefiting the general consumer trend towards Australian destinations – a Sydney respondent stated that 'Australia has caught up as far as client preference is concerned'. One Melbourne respondent portrayed Queensland marketing as being typified by the Gold Coast image: 'they advertise the Gold Coast more than anything – they should concentrate on the Whitsundays'. The larger capacity of the Gold Coast could be cited in justification of the present approach, however, the comment is an indication that the Queensland islands tend to be subordinate to other destinations such as the Gold and Sunshine Coasts and Cairns and that the issue of balanced coverage has not been entirely resolved. The disappointment of some clients in discovering that the Whitsundays did not provide a profusion of palm trees (the QTTC logo is a palm tree) received comment. According to one Melbourne agent, 'the composition of the islands is not like people's idea of an island – with de-vegetated areas without palm trees'. The fact that the Whitsundays are subtropical and less green and lush than the tropical Mamanuca islands, may be a marketing issue. The appropriation by Queensland of the 'classic' symbol of the South Pacific could be a negative for consumers who expect a lusher, more tropical environment. 'Lushness' is not a potential marketing platform in the Mamanucas either, because of their location in Fiji's dry zone.

The only respondents who stated that current promotion of the Mamanucas was sufficient were those who described the group as 'best kept the way they are' and preferred their existing quietness and seclusion. Even these respondents were not single-minded, since there was a clear perception that reinvestment in facilities is needed. Quietness and seclusion should lend themselves to niche-based marketing, though little is currently in evidence.

Fiji's consumer promotion was viewed as insufficient, too narrowly focused on resort facilities and inadequately backed by the FVB and by the individual resorts as far as agents were concerned. The main difference between Melbourne and Sydney respondents was the stronger Sydney dissatisfaction with flight times. Sydney's relative advantage as the Australian headquarters of both Air Pacific and the FVB did not appear to help. In Melbourne, Air Pacific is dependent on the Qantas sales office for all customer inquiries and reservations and the FVB office has only three staff. The criticisms of Fiji's marketing are not evidence that performance has always been inadequate – the generally positive disposition of travel agents towards the destination must have come about because of positive client experiences, presumably resulting from marketing effort. Nevertheless, promotional activity is clearly less potent than in the past.

In tourism marketing, *place* refers to two dimensions. Firstly it is the location of the product (in this case the two island groups and the individual resorts). Secondly it 'is the location of all the points of sale that provide prospective customers with access to tourist products' (Middleton 1994: 65). *Actual* accessibility is based largely on flying times from the major sources of business. These are outlined in table 5.7.

The Whitsundays clearly enjoy an accessibility advantage over international destinations. Fiji and Bali are the closest major tropical resort destinations for Australia. The Whitsundays claim the greatest advantage in Sydney, with the flight time to Fiji some 61 per cent longer than its equivalent to Hamilton Island. Hamilton Island resort itself is only minutes from the airport terminal. In Melbourne, the advantage is less pronounced with a 17 per cent difference in average flight duration. International travel does involve extra time for checking in (90 minutes versus 30 minutes) and for customs clearance. Certain Whitsunday islands expedite the check-in process by undertaking procedures on the launch prior to arrival on the island. Some of these differences may lead to a perception of much easier accessibility for the domestic destination. Nevertheless it is clear from a comparison with competitors such as Bali and Hawaii that as an international destination, Fiji is very accessible.

Sydney travel agents perceived the Whitsundays as easily accessible. The perceived inaccessibility of the Mamanucas is exacerbated by the inconvenient flight arrival times which were commented on by many Sydney respondents, who added that clients want to travel directly to the island and not stay overnight in Nadi. There is some concern about the accessibility of the outer Mamanuca Islands and about the safely of some launch transfers. Melbourne agents largely described access to Whitsundays as 'fairly direct'. Whilst there was some dissatisfaction about the flight times into Fiji (too early in the morning), the complaints were relatively less frequent and less insistent than in the case of Sydney. The easier accessibility of the Whitsundays is less pronounced for Melbourne holidaymakers.

Many respondents were reluctant to compare access to the two groups directly. Some categorised Fiji as international, implying that clients use an equivalent frame of reference. Viewed in this way, Fiji is the most accessible of all international sun, sea and sand destinations, though the issues of flight availability (less frequent than to the Whitsundays), scheduling (relatively inconvenient) and promotion (the range of available transfers to the islands

Table 5.7 Non-stop air service flight times between state/territory gateways and countries of intended stay (in hours and minutes)

City of departure	Destination airport			
	Hamilton Island	Nadi	Bali	Honolulu
Sydney	2hrs20mins	3hrs45mins	6hrs25mins	9hrs20mins
Melbourne	3hrs55mins	4hrs35mins	6hrs	NA

Source: ABC Flight Planner and ASMAL
Note: NA = Non-stop air services unavailable

registers low awareness) are also relevant. A lack of seat availability is also perceived as a problem for the Whitsundays, particularly the shortage of discounted seats during school holiday periods, apparently caused by the high demand and limited bed capacity at such times. Operators clearly benefit from higher yields during peak times, though the results of this research suggest that the region's image of affordability is affected negatively. The other negative is the inconvenience of transfers from Proserpine (a combination of coach and launch) and resentment that only one airline (Ansett) flies to Hamilton Island.

Most of the Mamanuca resorts are easily accessible from Nadi with Musket Cove and Plantation Resort a short ten-minute flight away. The price, frequency and short flight time are seen as advantages in Melbourne. Flights from Sydney arrive late at night and depart early in the morning, necessitating at least one and often two overnight stays in Nadi for NSW visitors. According to D. Smith (pers.comm.) the Civil Aviation Authority of Fiji (CAAF) cancelled the licence which previously allowed nighttime connecting charter flights to Malolo Lailai using lights on the airstrip. The requirement of a two-night stay in Nadi is a spectacular disincentive and the reasons for discontinuing charter flights merits reconsideration by CAAF in view of the crucial importance of accessibility for the Sydney market. The recent initiative of Air Pacific to service its aircraft in Nadi has altered the circumstances of flight connections since the earlier CAAF decision. In darkness, the only accessible islands are Treasure and Beachcomber which use a commuter boat for the two-mile crossing from Lautoka across a relatively placid stretch of water. A coach transfer is, however, required from Nadi to Lautoka.

Most Whitsunday resort managers perceive accessibility through Hamilton Island Airport as a major advantage when marketing within Australia (see figure 5.6). The two smaller resorts of Palm Bay Hideaway and Hook Island Wilderness Lodge are least well serviced (though both are readily accessible from the mainland). Palm Bay has always been the last stop-off point for launches from Hamilton Island, partly because of its small size (priority for islands frequented by bigger numbers) and partly because, until the 1994–5 programme, it was included only in the Australian Airlines/ Qantas brochures (accessed via the mainland) and not the Ansett Holidays programme.

The relatively short distance between the Whitsunday islands is a major marketing benefit. However, Brampton's only connection with the Whitsunday islands is a three or four times weekly cruise to Lindeman and to Hamilton. Brampton is serviced from Mackay and not Proserpine or Hamilton Island Airports. The resort is fortunate in having a more populous (and a relatively prosperous) hinterland around the city of Mackay and calls itself 'Mackay's own island', luring many locals for the weekend including residents from the coal-mining areas around Dysart (G. Mahony

Figure 5.6 Tropical paradise?
An airport in your backyard. Guests may welcome the easy accessibility, but not the aircraft noise. Did they not travel to 'get away from it all'?
Source: Gill Rawlings (courtesy of Ansett Australia)

pers.comm.). This advantage does not extend to the interstate travellers who are the subject of the present study. The latter must make a very conscious decision to visit the resort since the adjoining region has insufficient profile to act as a magnet. In the case of the eight *bona fide* Whitsunday islands, travellers may be persuaded that the region is their preferred destination because of its scenic and other attributes before determining which particular resort is best suited to their requirements. This implies a symbiotic relationship between the regional concept and the individual resorts. When promoted by its resort owners Qantas, Brampton places less emphasis on its regional credentials than do the Ansett-associated Whitsunday resorts. Given increased recognition of the Whitsunday name, this is to Brampton's disadvantage.

The close proximity of a number of medium- to large-scale resorts in the Whitsundays has created economies of scale to justify investment in inter-island craft and regular inter-island services. Whitsunday All Over vessels connect with all flights to and from Hamilton Island, offering much greater frequency than South Seas Cruises is able to offer with its single Island Express craft. Like South Seas Island Cruises, Whitsunday All Over has a monopoly of water transport connections from the airport, but a number of other Whitsunday companies such as Fantasea Cruises and Whitsunday

Connections offer competition in day cruising, including inter-island transfers. The quicker options which exist in the Mamanucas such as helicopters and seaplanes are costly, suggesting a lower level of accessibility except for the most affluent visitors. This is a dilemma of tourism in developing countries, where the absence of a domestic market makes the provision more marginal.

Fears were expressed by mainland operators when Hamilton Island Airport was opened in 1984 that the mainland would miss out on tourism growth. In fact, the creation of major airport infrastructure *within* the island group has provided an impetus and rationale for the projection of the destination names and image. Malolo Lailai Airport may only be a 10-minute flight from Nadi but its very separateness creates a perception that the Mamanucas are a group apart. In the absence of any active promotion displaying accessibility, the group will remain at a relative disadvantage. This reinforces the need for a resumption of night charters to Malolo Lailai Airport.

Relative to Sydney, part of the extra flight time from Melbourne to the Whitsundays is explained by the routing of flights via Brisbane or Sydney (direct flights are operated only during peak school-holiday periods). Vincent claimed that 'the lack of direct flights from Melbourne is not a problem (for Daydream) because there is usually only one stopover' (pers.comm.).

The dependence of the Mamanuca resorts on Australian-based tour operators is highest amongst the larger resorts for example, Mana, Castaway and Plantation. Matamanoa attributed 87 per cent of the Australian business at Matamanoa to 'big' tour operators (H. Steinocker pers.comm.). The smaller resorts, notably Navini and Tavarua, are less dependent – both have their own subsidiaries which act as travel agencies and take bookings in major source countries (Australia and the USA respectively). Sheraton Vomo, targeted at the 'deluxe end of the market', claims to work through a very small number of agencies and wholesalers.

The major tour wholesalers cited were Jetabout and Viva!, considered as virtually interchangeable by most resort operators. The other significant player cited was Rosie Tours, a Fiji-based operation. Headquartered in Nadi, Rosie's has offices and sales staff in Australia and is one of only two Fiji-dedicated based tour operators, Tapa Tours being the other. Rosie's has been active in the development of computer reservation systems and databases and in the promotion of special interest tours. The Mamanucas group could benefit from an even closer strategic relationship with Rosie's. A number of other operators were cited as significant by resort managers; these included Orient Pacific Holidays, Connections Holidays, Tapa, Swingaway Tours and Jetset. Such operators have dominated Mamanuca resort marketing because of the requirement that the cheapest airfares (tour-basing) form part of an overall package including accommodation. This specification has bolstered tour operators, but has created difficulties for resorts attempting to offer attractive rates to repeat customers as a reward for booking direct. As stated in chapter 4, repeat customers are particularly important to island resorts in Fiji. Apparently an 'official back-door' has existed for resorts to secure the cheaper fares but not on a significant scale. Until the recent loosening of the air-fare regulations, any discount offered by the resort could be negated by the need to pay the higher 'excursion' air fare.

The biggest problem created by the virtual stranglehold of tour operators is that resorts have little reason to do business directly with travel agents, nor to communicate with potential customers in the source markets. Travel agents and not tour operators have the major dealings with and hence understanding of customers. Both Musket Cove (D. Smith pers.comm.) and Castaway (G. Shaw pers.comm.) stated that they had recently introduced a procedure for clients to identify the name and address of their travel agent at the time of check-in. The fact that such an obvious procedure has come about so recently, is an indication that the long distribution channel is causing communication problems. The lack of consumer brand loyalty to tour operators in Australia exacerbates this problem (S. Brewster pers.comm.).

Information about the distribution methods used by the various Whitsunday resorts was difficult to secure. A number of resort managers were willing to criticise the airlines over the price of tickets and over flight frequency and capacity, but few were willing to criticise the structure of wholesaling to the resorts. This is despite – or perhaps because of – the fact that Ansett Holidays, a fully owned subsidiary of Ansett Australia, controls the vast bulk of air-based holiday packages to the islands.

Ansett's dominance is highlighted by its active promotion of the Whitsunday brand. In view of their ownership of Hayman and (until recently) their leases at Hook and South Molle and their investment in Hamilton Island Airport, the commitment is not surprising. One grievance with Ansett was expressed by Vincent. Referring to the Whitsundays All Over water-taxi service to Hamilton Island Airport, he stated that:

> Price is a problem and becoming worse. Ansett take a cut and then they take another cut and then they put on a fee for Ansamatic (the airline's computer reservation system) and then for baggage handling. The water taxi company don't come out with as big a chunk as people think. It's difficult for us. The price is going from $66 per person return to $74. That's one room night at rack rate.
>
> (pers.comm.)

Most resorts distribute through a large travel agent network. As part of its move to soften its image of exclusiveness, Hayman ceased its distribution through wholly-owned 'exclusive' sales outlets in Sydney's Double Bay and

Melbourne's Toorak and now distributes through all travel agents stocking Ansett and Sunlover brochures.

In contrast, South Molle deals mainly with Ansett-affiliated agents. Collins (pers.comm.) regards 'travel agent relationships as a major success factor for this property, probably more than is the case with the other resorts'. This strong affiliation with one airline group or the other (usually Ansett in the case of the Whitsundays) is symptomatic of the more polarised distribution network compared with Fiji. The competition for the Mamanucas is likely to intensify as the teething problems following the Qantas/Australian merger are resolved.

The other unusual distributor is Club Med, which chose not to be included in the Ansett, Qantas or Sunlover holiday brochures (until it was featured in the 1994–5 Ansett Holidays, indicative of a policy shift). The company distributes its own brochures to travel agents throughout Australia. Lindeman is the sole resort featured in its Club Med Australia brochure. The other main brochures are Club Med Pacific and Club Med Asia. Club Med's all-inclusive price allows it to operate a highly vertically-integrated operation. No other Whitsunday property has relied so predominantly on its own packaging and distribution.

Overall the distribution methods in the Whitsundays are highly concentrated in the hands of Ansett Holidays. Whilst Sunlover Holidays does allow for some diversity, there is no significant competition from independent operators. Qantas is unlikely to emerge as a serious competitor until further mainland development occurs.

Process involves procedures to involve customers in the process of service delivery and the deployment of computer reservation systems (Cowell 1984). This marketing mix element has been included in recognition of the increasing importance of automated techniques for resorts. It also allows one to assess whether Fiji (and the Mamanucas) have been marginalised through an absence of transnational investment in systems comparative to the QTTC's investment in the Australian tourism and leisure automated system (ATLAS) in Queensland.

Resort managers stated that the packaging of the Mamanucas in Australia is adequate. Coverage by the various computer reservations systems offered by the tour operators (for example, Jetset's Worldlink) were also considered adequate. None perceived the Mamanucas to be relatively disadvantaged. The reality, however, may be rather less positive than the perception. QTTC has successfully sought to persuade travel agents to book via ATLAS rather than telephone (S. Gregg pers.comm.). In contrast, tour operators featuring South Pacific destinations such as Orient Pacific Holidays have complained that agents are too dependent on telephoning tour operators to make basic inquiries easily answered by the various South Pacific manuals or by the computer reservations system (C. Willett pers.

comm.). The resort managers may be guilty of underestimating the relative advantage held by the bigger, more automated resorts in the Whitsundays.

A deficiency of on-island computerisation was acknowledged. Castaway, a medium to large resort, has a computerised accounting system but has nothing comparable to handle reservations, meaning that tour operators are unable to interface directly with current accommodation stock. Under pressure from tour operators to provide a seven-day 'release' for unbooked accommodation, the resorts are sometimes unable to sell their own rooms until a mere seven days before the room is due to be used. According to Geoff Shaw, rooms are unlikely to find a buyer at such a time. The Castaway manager suggested thirty or possibly fourteen days as a more realistic release period. Providing operators with immediate status reports on room availability would resolve the problem but is dependent on investment in systems. Mana Island is atypical in having benefited from lavish investment in systems by its Japanese owners. Its F$150,000 Micros system interfaces directly with head office. The resort manager also deploys the system to undertake statistical trend assessment, aimed at rewarding tour operators according to the volume of business generated. The Director of the Fiji Department of Tourism has acknowledged the lack of automation and has indicated his intention to seek investment allowances for properties intending to computerise (M. Gucake pers.comm.). His sudden death in late 1994 has placed a question mark over this initiative. The proposed development is clearly needed. The Mamanucas are generally leased to small to medium-sized businesses and do not enjoy the networking and distribution advantages available to international operators present in the Whitsundays such as Holiday Inn, Club Med and SPHC which have fully automated accounting and reservation systems.

Packaging may be defined as 'the combination of related and complementary services into a single-price offering' (Morrison 1989: 246). Most international air-inclusive packages are put together by tour operators, or wholesalers as they are sometimes called. Packaging can also be undertaken by travel agents or by airlines. Travel agent respondents down-played the significance of the tour operator with 84 per cent agreeing with the proposition that 'clients leave the choice of tour operator to the travel agent when booking these destinations'.

In contrast, Whitsunday resort managers regarded tour operators as performing a valuable marketing role. They viewed the colour brochures launched annually by the operators as a major outlay with the high production costs, incorporating colour photography and copy dedicated to particular resorts, seen as indicative of a major commitment to the destination. The fairly lavish style of airline brochures used are clearly a useful tool to bring about loyalty from resort managers.

Certain trends have constrained tour operator marketing, including a squeeze on margins, with the price of accommodation and airfares having risen relatively slowly. Unsurprisingly, the accommodation and airline sectors have different perceptions about who is responsible for this occurrence (the resorts operators say airfares are too high, Air Pacific accuse the resorts of excessive pricing). Air Pacific cited a return excursion airfare from Sydney to Nadi as having cost A$472.00 in April 1983, with the corresponding fare in January 1994 being A$656.00, a rise of 38 per cent in almost a decade (E. Dutta pers.comm.). The margin passed on to tour operators was cited as 13–15 per cent in Australia, considerably less than the figure for New Zealand (15–22 per cent); 9 per cent plus a possible override would need to be passed on to the travel agent. Clearly the profit margin available to tour operators in such circumstances is highly constrained. Despite its businesslike approach, Mana made 'substantial losses' in 1993 (K. Palise pers.comm.).

The prices offered by the resorts to Australian tour operators range between 30, 32 and 34 per cent off rack rates at Mana, dependent on the volume of business generated (K. Palise pers.comm.). Off-season discounting may provide operators up to 48 and 50 per cent off rack rates. According to Palise, such extreme discounting can only be sustained because the yields on other sources of business (mainly Europe and Asia) are much better. The implication is that current promotion to the Australian market is excessively price sensitive, with both operators and resorts under pressure. This cycle needs to be broken through a more strategic marketing approach with resorts placing greater reliance on their own devices.

The withdrawal of Qantas from the marketing of Fiji has placed an increased pressure on the marketing budgets of the tour operators. Other industry trends have included a move by consumers to shorter lead times and last-minute bookings. This 'deal' mentality has resulted in a change in the way full-price, full-colour tour operator brochures are used by consumers. Discounting has resulted in the production of a plethora of (usually) two-colour leaflets which promote special deals applicable for a limited period. These so-called 'flyers' offer either discounted accommodation and/or value added features such as 'free' breakfasts or 'learn to dive' packages. Other typical features are 'kids go free' or 'seven nights for the price of five'. The purpose of these sales promotion materials is to raise the prominence of Fiji resorts in a crowded marketplace. The result, however, may be confusion, since a complex array of validity dates, bonuses and discounts are on offer.

Jim Saukuru of Treasure Island (pers.comm.) stated that 'there may be too many flyers for the Mamanucas as a whole – at times this causes confusion for customers'. Most flyers have insufficient space to project logos, familiar colour schemes and other typical features which help induce brand familiarity and loyalty. The result is the opposite of rationalisation, in that as soon as one operator or resort introduces a special, others will do likewise, though usually with some extra complicating dimension attached. One manager stated: 'I have seen flyers stapled together that are current and are being used as scribble paper. It confuses the market' (M. MacDonald pers.comm.). Also of concern was the fact that a number of operators chose not to publish full-colour brochures for 1994, relying on a series of short-duration flyers instead. The result is a possible loss of product quality, at least as perceived by customers when exposed to the brochure range. The confusion may be compounded in the Mamanucas because of the plethora of tour operators in the marketplace, especially as each one appears to want to have its own distinguishing mark on the packages. In contrast, the control by the airlines and Sunlover Holidays in the Whitsundays has probably created a more stable promotional environment. Short-term initiatives are at least undertaken within an overall national framework.

Whitsunday resort managers viewed 'deep discounting' outside the main school holiday periods as a permanent feature of the region's marketing. Initially certain resorts, particularly Hamilton and Hayman, resisted pressure to reduce their rates at any time, continuing their emphasis on 'exclusivity'. More recently both resorts have discounted their prices during off-peak periods. Hamilton was the last to do so, having emphasised a 'value added' approach until late 1993. Since 1994, the rates offered by Hamilton as part of Ansett's *Great Deals* programme are pitched slightly under those charged by Club Crocodile (formerly Radisson) Long Island Resort which previously offered the lowest rates. Daydream is typical in charging rack rates during school holidays.

Despite the substantial occupancy hikes evident in the Whitsundays, the deep discounting has an impact on the per capita spending of Australian visitors. At Brampton, Mahony (pers.comm.) said that 'we have attracted [with $49 per person accommodation twin share] more budget travellers who could not have afforded us before. We have experienced a downturn in expenditure on food and beverage and in the boutiques.' Clearly a balance is needed between higher-yield visitors and top-up lower-spending customers who can help to finance the high cost of resort operations during low demand periods.

The introduction of attractive lead-in prices has brought promotional benefits to the Whitsundays. Ansett has advertised its Great Deals very widely in the metropolitan media under the Whitsundays label. This should raise consumer awareness of the discounted holidays available and overcome the predominant perception of inaccessibility and exclusivity. Though advertising less prominently, Sunlover's equivalent 'Dollarwise' packages helped raise the area's profile.

The potential desirability of the Mamanucas as a short-break destination was outlined at Sheraton Vomo. A feasibility study for the resort conducted by the World Bank anticipated an average length of stay of three-and-a-half to four nights, an indication of the potential of the short-breaks market. During the months after the resort opening, Vomo attracted many long-weekend holidaymakers out of New Zealand and anticipates more from Australia. 'Three- and four-night stays will definitely become a strong proposition out of Australasia as the market settles down' (N. Palmer pers.comm.).

The TCSP Marketing Head (S. Lolohea) was adamant that the Mamanucas could tap the affluent 'dink' (double income, no kids) short-break travellers out of Australia. This view was shared by the Chief Executive of the Fiji Hotels Association (R. Singh pers.comm.), who saw high spenders on short breaks as very desirable. The view was not shared by the FVB, who saw any further reduction in the length of stay (B. Whiting pers. comm.) as undesirable. Whiting also claimed that the short breaks had been targeted by New Caledonia out of Australia with little success.

Other managers did not talk about a potential growth in short breaks but, like the FVB, saw an undesirable trend towards a shorter average length of stay. The other impediment is the airfares, which have a seven-day requirement for tour-basing and a five-day requirement for excursion fares. A plethora of three-night Whitsunday packages ex-Melbourne and ex-Sydney are now being offered by Qantas, Ansett and Sunlover Holidays, though they are regarded by resort managers as a 'marketing ploy' in that most customers responding to such promotional initiatives chose to add extra nights to the minimum duration (S. Brewster pers.comm.). Nevertheless the perceived ease of access implied by such short breaks is a valuable means of overcoming perceptual barriers about travel time and the feasibility of achieving relaxation on an international holiday when time is very short.

A marketing opportunity is signalled by the recent success of Air New Zealand's packaging and marketing of short breaks from Australia to all major cities and tourism destinations across the Tasman (Marshman 1995: 8). The Air New Zealand experience suggests that the fashion for international short-break travel so evident in Europe need not be confined to the Northern Hemisphere. Fiji's easy accessibility relative to its South-East Asian competitors suggests that a key marketing advantage could be secured if key decision-makers at the FVB were to be more responsive.

FUTURE MARKETING PRIORITIES

Travel agents were asked to specify issues requiring attention by the Whitsunday and Mamanuca authorities. Table 5.8 lists the responses. The percentages apply to valid responses only and do not take non-respondents into account.

For travel agents, the major marketing priority for Queensland regions such as the Whitsundays should be price, confirming the findings of the

Table 5.8 What should the Queensland and Fiji tourism authorities be doing better? (Priorities indicated in parentheses)

Action	Queensland	Fiji
Prices	26.9% (1)	10.0% (4)
More detailed information	21.0% (2)	30.0% (1)
Advertising	15.1% (3)	26.5% (2)
Public awareness	9.2% (5)	11.8% (3)
Good packages	10.1% (4)	9.4% (5)
More incentives	9.2% (5)	5.3% (6)
Keep updated	5.0% (7)	3.5% (7)
Other (bonuses/properties)	3.4%	3.6%

Source: Travel agent survey, December 1993
Note: Missing frequencies: Fiji = 30; Queensland = 81
        n = 200

SWOT analysis. For both regions more detailed information was demanded, consistent with a consumer desire for better understanding of their holiday destination at regional as well as at state or national level. One might argue that tourism promotion agencies can never hope to satisfy the insatiable demand for more information. Nevertheless the method of information dissemination deserves some reappraisal by both destinations. Enhanced advertising to improve public awareness is urged for the Fiji authorities.

Undertaking a direct comparison between the effectiveness of the marketing of the Whitsundays and of the Mamanucas is hampered by a lack of directly comparable data. The absence of data on regional and resort promotional expenditures (as opposed to state and national expenditures) is an obvious limitation and means that attempting to draw direct correlations between promotional expenditures and visitor numbers is risky. We must also be aware of the different resource bases in a developed country (Australia) and a developing country (Fiji). Finally, the importance of external factors such as the changing role of Qantas reduces the validity about assertions concerning marketing effectiveness.

It is clear that promotional expenditure by both the Queensland authorities and the Whitsundays region far outstrips the comparative figures for Fiji and the Mamanucas. The greater cohesion of the Whitsundays as a marketing unit and the support provided to it by the QTTC have resulted in greater consistency of promotional messages. The Fijian and Mamanucas approach has been less consistent, with short bursts of (costly) television advertising and sudden changes to the airfare regime. Overall, agents regarded Fiji's fatal flaw as marketing, rather than product quality.

The contrasting approaches to strategic regional marketing are striking. The Whitsundays work to a business plan incorporating detailed assessment of SWOTs and dovetails with the Draft Whitsundays Strategic Tourism Plan (WVB 1994). No strategic marketing plan is in evidence

141

for the Mamanucas. Even the FVB Marketing Plan (1993a) is notably stronger on promotional tactics than on strategic objectives. The section of the current document which deals with SWOTs is weak on detail. The failure of the 1989 Fiji Tourism Masterplan was a major shortcoming, since rejection of that document by industry coincided with increased deregulation of the economy and the abandonment of the national planning system (Development Plan 9 concluded its five-year span in 1990). Whilst the relative flimsiness of the FVB Marketing Plan might not in itself retard tourism development, the fact that it operates in a policy vacuum has significant consequences. The absence of a lobby group representing all of the tourism industry (as opposed to separate sectional representation for the tour operating, accommodation and duty free industries) has a further impact.

In the Whitsundays, change has resulted from good fortune as much as good planning. Commonwealth involvement through the GBRMPA and through the process of World Heritage Listing prevented the *laissez-faire* attitude of the Bjelke-Peterson Government from corrupting environmental standards. The natural and climatic attractions of the region were largely responsible for the willingness of major international hotel groups to take on the management of island resort properties and to apply their consistent service standards. The genuine and consistent commitment of the QTTC to regional tourism promotion was one example of good planning.

This chapter has conclusively identified a greater awareness of the Whitsundays to that of the Mamanucas in Melbourne and Sydney. The Whitsundays have benefited from the increasing propensity of Australians to take air-based interstate trips, a desire fuelled by Ansett Australia's commitment to promoting the Whitsundays name and by the reduction in travel costs following deregulation. Gains made by the Whitsundays within NSW and Victoria have not been made at the expense of Fiji alone, but they have undoubtedly had an adverse impact on travel to Fiji.

In this section of the book our focus was on the tourism industry and away from the islands themselves. In Part III our focus reverts to the islands under the heading 'the raw and the cooked', an expression borrowed from the French anthropologist Claude Levi-Strauss (1970). In the present context we refer to the 'raw' as the social and environmental state of affairs prevailing in pre-industrial times. The 'cooked' refers to the built environment, to social engineering and to manufactured images which are more symptomatic of the industrial or post-industrial period. Holiday resorts have the effect of 'cooking' the whole environment. Too much cooking is probably a bad thing, since a variety of (albeit more synthetic) alternatives to island resorts are emerging and will force the islands to reappraise what unique benefits they can offer to their clients.

Part III

THE RAW AND THE COOKED

6

FITTING THE IMAGE?

THE MYTHICAL SOUTH PACIFIC: IMAGERY SINCE THE TIME OF THE COOK VOYAGES

In chapter 1, the long historical pedigree of the 'earthly paradise' was identified. It was shown that tourism promoters exploited the earlier literary tradition by arguing that the fictional dream could become reality for European consumers. Such promoters played down the fears of solitude and savagery underlying many literary accounts and emphasised the positives. Whilst any subtropical or tropical island destination has the opportunity to tap the imagery of 'the classic tropical island', it is the South Pacific which most unequivocally is thought of as the 'classic' location for such fantasies of the Western imagination. There is, however, an ambiguity which possibly is most evident in the case of Fiji.

Cohen (1982) has distinguished between the paradise myth of the Polynesian coral islands and 'the less widespread but powerful image of uncouth, primitive wilderness' which 'conjures up the themes of primeval savagery and cannibalism, of uncivilised man and untamed nature, of mysterious dangers lurking in dark jungle forests' (Cohen 1982: 13). Such images are most closely associated with Papua New Guinea, though they also apply to the other Melanesian countries of Vanuatu (with its volcanoes, though it was also home to Michener's *Bali Hai*), the Solomons and New Caledonia. Underlying this imagery is the view that Melanesians exhibit the characteristics of 'hard primitivism' and Polynesians the more appealing 'soft primitivism'. Although Fiji is a 'bridge' between Polynesia and Melanesia with some Fijians displaying the lighter Polynesian complexion and others the darker Melanesian, the Melanesian is the predominant strain and Fiji is typically described as a Melanesian country. The tourism imagery of Fiji relies more heavily on symbols associated with Polynesia, such as photographs of palm-fringed coral atolls. According to Brookfield and Hart (1971: 378), tourism imagery has brought about a gradual 'pseudo-Polynesation of Melanesian Fijian culture'. The Mamanucas are a notable example of the Polynesian style imagery since they are principally low-lying

145

coral atolls rather than rainforest clad, mountainous and volcanic. To date the focus of Fiji's tourism marketing has emphasised island and coastal resorts. This is now changing as the FVB and the Fiji Department of Tourism have shifted the emphasis to ecotourism and adventure tourism. Such activities (whitewater rafting and jetboating are examples) are mainly undertaken in the inland and more mountainous parts of Fiji. This invites the questions of whether Fiji will become increasingly influenced by consumer perceptions of Melanesian imagery. Will such imagery complement the existing (Polynesian style) imagery as is obviously intended by the authorities, or could it undermine the image of Fiji as a 'South Seas paradise'?

One dimension of Polynesian imagery which Fiji has avoided has been the conscious portrayal of female sexuality and sometimes implied promiscuity that has been the hallmark of promotion of French Polynesia in particular. Brochures for Fiji dating from the 1920s and 1930s depicted topless Fijian women, though they were portrayed more as objects of interest than as subjects of desire. Since the Second World War, the dominant depiction of Fijians (rarely Indians) has been as fully dressed, friendly and smiling. In this respect at least, Fijians with their darker, more Melanesian appearance do not fit the Gauguin depiction of Polynesian women with its associations of sexual attractiveness and availability. The change over time is probably attributable to the increasing shift away from eighteenth and nineteenth century depictions of the 'noble savage', though one should note that the absence of sexual connotations has been a constant in the tourism promotions for Fijians, possibly a result of the strong missionary influence.

From the vantage point of the 1990s, any claim by the Whitsundays to be 'classic South Pacific' would be regarded as more questionable. Australia is more typically depicted as a 'part of Asia' than as part of the South Pacific. The message propagated by politicians and business leaders is that Australia should emulate the economic success and fast growth of the Asian region. In contrast, the countries of the South Pacific have experienced little or no economic growth over the past decade and are often regarded as an economic and political backwater. In light of these facts, Australian politicians typically would prefer to avoid consciously emulating the South Pacific. The absence of an established indigenous population in the Whitsundays also makes parallels between the two tropical island groups seem far-fetched.

Although there is a lack of convergence between tropical Australia and the islands of the South Pacific from an economic and geopolitical point of view, some other trends have drawn the two closer together. The recognition of Aboriginal native title by the Australian High Court's 'Mabo Decision' (1992) subsequently reinforced by Commonwealth Government legislation has eliminated one of the key contrasts between Australia and its immediate South Pacific neighbours in the 1990s. The Pacific islands are regarded as being ahead of Australia in recognising native landownership. The tendency for issues of environmental management to become internationalised has also drawn Australia and Pacific countries such as Fiji closer together. Former Prime Minister Keating's strong stance on environmental management issues at meetings of the South Pacific Forum, particularly the 1994 Brisbane meeting, is an indication of Australia's growing interest in the field. Australia has depicted itself as sharing common interests with the South Pacific nations against the predatory commercial activities of 'outsiders' from Malaysia, Japan and Taiwan. Increasing environmental awareness is of particular relevance for both the Mamanucas and the Whitsundays because the quality of the natural environment is the key attraction offered in both regions.

Convergence of imagery is also being prompted by the rapidity with which Australians are reconceptualising their own tropical north. The early white settlers of Australia favoured the more temperate climes prevailing in the southern States which reminded them of 'home' in Europe. Over time, the tropical north has developed its own appeal and mythology within the growing Australian population. Altering the perception that the tropics are a place unsuited to European manual activity was certainly a slow process. The use of low-wage Asian or Pacific Island labour which had underpinned the early plantation economy of North Queensland ended with the adoption of a 'White Australia' policy following Federation. Gradually the realities of life for whites in the tropics were improved with medical advances and new technologies, a point noted by Breinl and Young (1920) and Cilento (1923). These authors claimed that tropical Australia did not experience higher white mortality, as was previously supposed, nor physiological changes, nor medical disabilities, except those accounted for by the inferior living and working conditions of the pioneers (Holmes 1985). By the late 1950s and early 1960s, the tropics were increasingly regarded as symbolising the benefits of life in the 'Sunshine State' of Queensland (though it was not until 1977 that the expression 'Sunshine State' was used on Queensland vehicle number plates) and the essence of statehood, though the bulk of postwar internal migration has benefited the subtropical coasts near Brisbane more than the tropical north. Cilento and Lack's *Triumph in the Tropics* (1959) opened with the words: 'The story of the self-governing State of Queensland is essentially the record of the white man's triumph over climate and his taming of the tropics'. This demonstrates that North Queensland was increasingly seen as 'our own', just as the Pacific islands were seen as a holiday playground for Australians.

The warmth of the north appealed as a remedy for illnesses associated with cool, damp climates (in novels such as *Travelling North*) and as a challenging frontier country less hostile than the arid 'outback'. The imagery used in tourism promotions dating from the 1930s to portray Queen-

147

sland's tropical islands was significantly different from those depicting mainland Australia. The vastness of mainland tropical North made it distinctly Australian, resistant to direct European parallels and an example of 'otherness'. In contrast the small dimensions of the Queensland islands were a reminder of the more human scale of Europe, a continent also surrounded by islands (albeit very cold ones for the most part!). As islands, the Whitsundays and other Queensland islands were relatively inaccessible from the southern population centres particularly prior to the arrival of jet aircraft. To the resident of Melbourne or Hobart, the Queensland islands would have seemed quite foreign. This foreignness was a reality prior to Federation when Queensland was a separate political entity from the other Australian colonies. In the late nineteenth century the Whitsundays appealed as a landscape that Australians could relate to, but with the appeal of physical impenetrability.

The Queensland islands are not 'classic' South Pacific, though the nineteenth and early twentieth century depictions borrowed from the imagery already associated with the (then) colonies of the South Pacific – Polynesia and Melanesia (see figure 6.1). Terms such as 'tropical paradise', 'my tropic isle' and 'isles of enchantment' were soon in circulation. The parallels with the South Pacific tropical islands were obvious. Many were uninhabited (hence 'deserted'), had sandy or coral beaches (hence the imagery of 'coral isles'), some were lush (hence the image of 'bountiful') and enjoyed a tropical climate (hence 'tropic isles'), were well endowed with potential foods (fruits and fish) and were protected by a reef (the Great Barrier Reef). There were of course many differences. The present-day Queensland island resorts are predominantly located on continental islands. Their flora and fauna is typically Australian and hence distinct from what is found anywhere else in the world including the islands of the South Pacific. Eucalypts, for example, constitute the predominant tree coverage and not palms. The indigenous population were distinct (Aborigines rather than Melanesians or Polynesians) and many islands were actually dry and not at all lush. Using the distinction between islands and 'continental' islands used by the French economic geographer P. Defert (1988), the Whitsundays are clearly an example of the latter being located on the continental shelf. Tourism promotion has, perhaps deliberately, linked the Whitsundays with the Great Barrier Reef, thereby implying a location further from the continent than in reality.

Discussion of the extent of differences and similarities between the Whitsundays and the Pacific islands is unlikely to arrive at a conclusive outcome. The 'classic' tropical Pacific island was always no more than a myth. It created an archetype or ideal out of diverse features from many different islands. Typically, it ignored the fact that some islands are flat and waterless coral atolls whilst others are volcanic mountain ranges rising precipitously from the sea, that many have few sandy beaches and that

Figure 6.1 Islands of the sun
This 1930s Queensland government brochure depicts islands offering classic South Pacific bures.
Source: 1930s Queensland Government Brochure

cyclones render the climate less than perfect. The diverse ethnicity of the inhabitants of the various islands was also underplayed.

To what extent did descriptions of the Queensland islands appropriate notions of the tropical paradise? Dunk Island became the archetypical Australian tropical retreat thanks to the writings of E.J.Banfield. His book *My Tropical Isle* was first published in 1911, its success leading to publication of two subsequent editions prior to the Great War. His work *The Confessions of a Beachcomber* (1908) was first published in London and reissued four times prior to the first Australian edition in 1933. Banfield's later works included *Tropic Days* (1918) and *Last Leaves from Dunk Island* (1925). He also influenced perceptions of the Queensland islands in Europe including the work *My Island of Dreams* (undated) by a Londoner, J.W. Frings. Frings described his favoured island as 'my Pacific Paradise'. The allure of the classic Queensland tropical island was enhanced by the rapidly developing literature on the Great Barrier Reef. Works by Roughley (1937) and Lock (1955, 1956) created an association between the allure of the Queensland islands and of the Reef, even though groups such as the Whitsundays are fairly distant from the natural wonder itself. More recent studies have shown the enduring and growing attraction of the Great Barrier Reef as an attraction (Vanclay 1988). Most of the nineteenth- and early twentieth-century accounts of the islands described encounters between settlers and the Aboriginal population in detail. The second half of *Confessions of a Beachcomber* (Banfield 1908), for example, is entitled 'Stone Age Folks'. Reference to such interactions is virtually non-existent in current promotional images of the Queensland islands. The early accounts portraying the Queensland islands as a 'perfect place to live out one's days' are similar to accounts of the imagery of the South Pacific, though a detailed reading of the literature brings out the diversity of landscape and culture in different islands. Stereotyping is certainly less blatant than in today's island promotions.

To understand the comparative imagery of the Fiji and Queensland islands, it is useful to start with the early exploration of the areas by European adventurers. The art historian Bernard Smith (1989, 1992) has argued that the three Cook expeditions and the voyages of other Europeans between 1768 and 1850 were responsible for transmitting images indicative of the preoccupations and dichotomies of the period. These included the debates over evolution versus creationism, 'soft' versus 'hard' primitivism, and imperial and commercial ambitions versus concerns about exploiting less technologically advanced societies. Smith has argued that the technical specialists such as draughtsmen, painters, botanists, ethnographers and diarists who played a significant role in realising the scientific and logistic objectives of the voyages were important purveyors and shapers of the subsequent Pacific imagery. These scientists, artists and technicians depicted the natural and the human environments of both the South Pacific and Australia. Though their impressions sprang from a European viewpoint, they provide us with valuable baseline data – both subjective and objective – for our two case study areas at a particular time in their development. The auspicing of the exploration of the Pacific by the Royal Academy placed the Pacific 'within the realm of non-civilised Natural Science: even the Noble Savage was so by virtue of his close relationship with Nature' (Sharrad 1990: 598). It is interesting to note that the growing interest in Aboriginal tourism in Australia closely parallels the rise of the environmental movement. The view of Aborigines as being 'at one with nature' is very convenient as a promotional platform to highlight the natural beauty of some Australian destinations.

Fiji and the Queensland islands were explored by Europeans just as newly pronounced scientific principles (such as Darwin's Theory of Evolution) were shaking many of the canons of European thought. Because many of the new theories went on to form the basis of much twentieth-century rational secular thought, the exploration and subsequent colonisation of the South Pacific had a greater impact on Europe than did equivalent explorations in other parts of the world. The exotic flora, fauna and ethnography of the South Pacific raised some awkward questions about previously accepted 'global' principles and the area became a key testing ground for the newly emerging scientifically-based world view (much of Darwin's experimental activity was undertaken in the Pacific). Europeans showed an avid interest in Pacific images which were often a curious mixture of rational and romantic elements. This dichotomy has formed the basis of more recent imagery. A variety of authors (for example, R.A. Britton 1979) have also stressed the continuity between colonial and postcolonial images of the Pacific. The continuity of economic, social and political linkages between the colonial period and the period of tourism expansion in the South Pacific has also been commented on by Steve Britton (1982; 1983) and by Britton and Clarke (1987).

The link between the emerging imagery of the eighteenth and nineteenth centuries and subsequent political and economic relations is argued by Smith who states that, 'The European control of the world required a landscape practice that could first survey and describe, then evoke in new settlers an emotional engagement with the land that they had alienated from its aboriginal occupants' (1989: ix).

One might argue that current tourism imagery perpetuates such attitudes by 'taming' the landscape for the enjoyment of international visitors (exclusive enjoyment in the case of uninhabited islands aimed at an overseas clientele).

THE TRIUMPH OF THE VISUAL: ADVERTISING AS THE LANGUAGE OF CONSUMPTION

Tourism imagery has been a popular subject of research amongst social scientists and amongst business studies researchers. Boorstin (1964) in his pioneering work *The Image* examined the impact of image dissemination on society generally. In one chapter 'From Traveller to Tourist. The Lost Art of Travel', he highlighted what he saw as the increasing predominance of 'gloss' over 'substance'. P.L. Pearce (1988) has described image as a term with 'vague and shifting meanings' and pointed to the wide range of concepts to which it had been linked, notably consumerism, attitudes, memories, cognitive maps and expectations (Lynch 1960).

Despite his warnings about the potential for misuse of the term 'image', Pearce subsequently urges tourism researchers 'to note the commonest meanings and connotations of the expression and adhere to these nuances of meaning' (1988: 162). He then points out three connotations of the word image. All of these are useful for the present research. Firstly there is a connotation with a publicly-held stereotype which is seen as embodying both people and place. In this sense the image of a resort would be inseparable from perceptions of the type of people that frequent it, or images of Fiji would not be imaginable without impressions of those who live there (a complex issue in itself because Fiji is described as 'multi-cultural'). Secondly Phillip Pearce (1988) points to 'a strong visual concept . . . a search of the long term memory store for scenes and symbols, panoramas and people' (1988: 263). Finally he observes that images are typically contrasted with reality. The last of the three is often associated with perceptions of authenticity, an issue which will be taken up later in this chapter.

Phillip Pearce (1982) demonstrated that perceived image exerts a significant and fundamental influence over leisure travel decision-making. A variety of researchers (Hunt 1975; Goodrich 1978; Crompton 1979) have shown that clear, emphatic and unthreatening images are those most sought after by consumers. Hunt's research offers insights into why we might expect Australian travellers to have a clearer image of a domestic destination such as the Whitsundays as opposed to an international destination such as Fiji. His study of the Rocky Mountains in North America concluded that respondents living further away from the destination area were less well able to identify the attributes of sub-regions than respondents living closer to the area. This principle leads us to expect that Australian travellers would possess a less distinct awareness of the attributes of the Mamanucas than of the Whitsundays. In the conclusion to his study, Hunt states that perception defeats reality where image projection influences potential consumers more than the actual recreational resources offered by the destination. Other studies have indicated that images change over time (Gartner 1986) and vary according to the consumer's previous travel experience (P.L. Pearce 1977; 1982). Cost, climate and scenery were shown to be the major factors in bringing about a positive consumer image for a particular destination in a study by Anderssen and Colberg (1973). Gartner

(1993) recently attempted to summarise some of the key characteristics of destination image. He hypothesised that:

1 The larger the entity (that is, destination), the more slowly images change;
2 Induced image formation attempts must be focused and long term;
3 The smaller the entity in relation to the whole, the less of a chance to develop an independent image; and
4 Effective image change depends on an assessment of presently held tourism images.

If we apply these hypotheses to the Mamanucas and to the Whitsundays we could expect that, as relatively small entities, the two groups could potentially alter their images more quickly than larger destinations, that their smallness relative to the whole (Fiji and Queensland respectively) limits the opportunity to develop an image genuinely independent of the larger entity and that change is unlikely without an improved understanding of current imagery. The present study aims to enhance this understanding with particular attention paid to the perceptions of consumers and of travel agents.

In the plentiful sociological literature, the French are particularly critical of what they perceive as the corrosive influence of modern 'promotional culture' on society as a whole. French critics regard the prevailing promotional culture as an unproductive distraction from the real concerns of society. Some of these criticisms have been applied to tourism through what is claimed to be a perpetuation of myths about tropical 'paradises' by tourism promotion. They have argued that the real nature of existence in the destination (particularly from the perspective of the indigenous population) is deliberately obscured with a view to ensuring the perpetuation of images which reinforce established consumer stereotypes. Cazes (1987: 37) regards the 'success' of tropical islands for tourism as based on what he describes as 'appropriation', 'insularisation' and 'mythification'. The relevant islands become 'possessed and dominated' by outsiders, insularised by the creation of 'enclaves and isolation' and mythified by the 'deliberate blanking out of social realities and the trivialisation of local culture'.

The sense of 'appropriation' was touched upon by the manager of Mana Island in the Mamanucas (K. Palise pers.comm.). He suggested that Australians had previously thought of Fiji as 'their own backyard', though he acknowledged that this view was eroding because of the increasing international spread of visitors to Fiji. A change of attitudes was coming about, however, as Australian holidaymakers acknowledged that holidaymakers from other sources were outnumbering them. The term 'appropriation' can still be applied to holidaymakers from metropolitan countries in general, though no individual nationality yet enjoys the virtual monopoly over such activity that Australia previously had. Cazes (1976) has argued that resort promotion builds upon a relationship of appropriation since islands are often depicted as 'belonging to tourists'. This in turn demeans the local culture and social structure. In the case of the Whitsundays, the absence of an indigenous population means that such attitudes are likely to have a less immediate impact, since there are few locals to be offended. One could argue, however, that the absence of references to Aboriginal settlement and/or incursions is evidence of the perpetuation of a sort of commercial *Terra Nullius*. As outlined in chapter 2, only one of the Mamanucas (Malolo Lailai) is not designated as native title. Indigenous ownership of land and control over the uses to which it can be put constrain the ability of resort developers to engage in appropriation of an explicit nature. Promotional images may nevertheless encourage holidaymakers to believe that they have 'purchased' possession and in a sense ownership of the island for the duration of their holiday. The absence of Fijians as 'customers' of the resorts can also lead to an impression that the land is set aside for the exclusive enjoyment of outsiders.

Cazes (1987) claims that of the sixty world destinations where tourism receipts are highest relative to total national revenue, the top fifteen are islands. Such evidence highlights the level of dependency on tourism experienced by tropical islands which usually possess few other economically tradeable assets (Hall 1994b). According to Cazes (1987: 37), the imaginative power evoked by tropical islands is 'formidable', a view reinforced by Bachelard's (1957) description of islands as possessing 'a powerful initial image' and as 'dominated worlds'. The tropical islands are the 'embodiment of all of our hopes' according to Decoin. Eliade's influential *Images et Symboles* (1952) argued that 150 years of European literature have created 'a sense of envy for the tropical paradise in the Great Ocean, refuges for all of our happiness, even though the reality is very different'. He has argued that the imagery of tropical islands has taken on theological proportions in an increasingly secular world. 'It had received, assimilated and readapted all of the paradise images refounded by positivism and scientism.' Eliade regarded the imagery of such islands as portraying them as outside time and outside history where 'man was happy, free and unconditioned; he didn't have to work for a living; the women were beautiful and eternally young; the perfect condition of man, of Adam before the fall'. Images take on a particular potency for resort islands where tourism is the sole economic activity and have the potential to overwhelm reality for the resort guest whose encounters consist of anticipation, a (usually short) stay, followed by reflection and reassessment. The reality of the experience becomes dominated by the anticipation and then by the recollection, captured in photographic form. The constant promotion of tropical island resorts experienced prior to and after travel creates a context or point of reference.

What one might call the 'distortion of reality' argument is taken up by several authors. Cazes (1987: 52) claims that tourism promotion has taken the paradise image and has 'vulgarised' it by 'brightening it up'. Cohen (1982) described the Pacific islands as 'a multitude of artificial and commercial paradises'. Cazes (1987: 52) likens the shifting back and forth between image and reality to a type of 'occult'. Equally critical of the approach of tourism promotion is Barthes (1957) who has stated that 'to apply colour to the world is always a means of denying it'.

The concept of self-discovery in a secure environment is another common theme in the literature. Paraphrasing R.M. Rilke, Bachelard (1957) proposes that holidaymakers can derive benefits from 'getting to know oneself in a narrow space' (interestingly, getting to know oneself was selected as the marketing pitch for 1994 by Brampton Island) (G. Mahony pers.comm.). An island is seen as a suitable place to return into oneself, to get back to one's source and to find intimacy with others in a small village setting. The individual is protected in a secure place, in the knowledge that difficulties and/or cost of access filter and discriminate between those who will be there. In this sense, a closed exclusive community is created. This has been the tenor of much Club Med promotion throughout the world (though the company preached equality and freedom as the context), and of Sanctuary Cove on the Gold Coast developed by Mike Gore (M. Gore quoted in King and Hyde 1989b), which was depicted as an oasis in a sick world. According to Bachelard (1957), 'the antenna of comfort and security bathes all messages, both textual and iconic'.

Eliade (1952) has described the self-delusion likely to eventuate from such an approach. The perfect image could be dispelled by the geographical reality of certain destinations and the image that everyone is slim and beautiful could be tainted by the presence of the fat and the ugly. The latter statement suggests that travellers are typically accompanied by the emotional 'baggage' of their normal lives in the metropolis. It also implies that island resorts raise excessive expectations amongst prospective guests, a view strongly denied by the General Manager of Daydream Island (B.Vincent pers.comm.) who claimed that most visitors to his resort have their expectations exceeded. Many businesses including resorts are very conscious of the pitfalls of inflated guest expectations and it is highly questionable whether professionally-managed resorts practise such exaggeration consciously. Nevertheless, the overall rhetoric of resort promotion as a *whole* may have the effect alluded to by Eliade.

The fantasy and 'unreality' of island resort imagery can be reinforced by certain styles of holiday organisation and by the physical properties of resorts. Cazes (1987: 43) talks about the development of 'pseudo-islands' which often use names such as village, holiday club, hotel complex, marina, condominium, leisure park and integrated resort. He claims that it is the very resorts which describe themselves as 'integrated' which are deliberately most physically isolated from their natural environs. Such islands pursue what Cazes (1987: 43) describes as 'multifunctionality and autarky'. He refers to the rapid growth of holiday organisations such as 'active discovery clubs' and 'formula clubs' as offering 'incantatory affirmations'. In both the Mamanucas and the Whitsundays, the 'Club' idea has been promoted. Club Med (on Lindeman Island) and Club Crocodile (on Long Island) are obvious examples of the *genre*. Naitasi Resort in the Mamanucas was previously a club when it traded under the name 'Club' Naitasi. A number of tour operators also promote the 'Club' idea. Orient Pacific Holidays offer their customers a *Passport Club* concept which offers benefits in return for a joining fee. Members of the Club are offered different, though roughly equivalent (in dollar terms) benefits by different resorts. Marketeers might argue that such concepts are typical methods used in all consumer markets, but the fact that resorts guests live together in a community over an extended period gives an additional social dimension to the Club concept.

Destinations offering even greater artifice may emerge as direct competitors for tropical holiday islands. Cazes (1987: 44) has pointed to the Club Med Austro-Tropical Village outside Vienna as an example of the 'tropics under glass' concept typically located on the periphery of major urban areas. This is an extreme example of a delocalised insular theme, though similar concepts are emerging as competitors for coastal resorts in Australia.

Cazes uses the example of the move by Hyatt into the resort business in 1987 to illustrate the increased emphasis being placed on resort architecture and facilities over the natural island environment. He states that 'with their unusual and spectacular architecture, they are becoming the objective of a holiday, even ahead of the destination in which they are found' (interview with the President of Hyatt in L'Echo Touristique on 23.3.87 quoted in Cazes 1987). In contrast to such architectural *gigantisme*, a number of Australian resort architects have deliberately planned island resorts as low-key developments where resort facilities are blended with the natural environment. The architects and designers of the up-market Dunk Island in Far North Queensland minimised the use of air-conditioning in favour of ceiling fans and open lattice screens, drew inspiration from the 'timber and tin' vernacular of North Queensland 'tempered in the main building by the great timber tradition of Japanese architecture' (Gazzard 1985: 67). The guest rooms were dispersed throughout the site, rather than clustered together in a large building. The success of Dunk Island (it reports a high occupancy level and good balance between international and domestic guests) indicates that there is an alternative to awe-inspiring architecture in the tropical island setting, though the opportunity may have been missed in the case of Hamilton, Daydream and Hayman (as depicted in figure 6.2) Islands.

Coastal settings close to urban areas in the tropical zone such as Singapore also pose a threat. A recent article outlined the rapid expansion of resorts around the city state. 'Does Singapore have what it takes to compete with Asia's purely resort destinations? The Singapore Tourist Promotion

Figure 6.2 Acres of swimming pools at Hayman
Hayman's vast swimming pool seems to make sea-bathing unnecessary and duplicates what the natural environment has to offer.
Source: Gill Rawlings (courtesy of Ansett Australia)

Board thinks so' (See Tho 1994: 7). Sentosa Island, located a few minutes from the Central Business District of Singapore, is. one example. The two resorts on Sentosa Island are the Shangri-La and the Beaufort. Both offer similar facilities to traditional beach resorts. Shangri-La also features an artificial beach. The heavily manicured landscape is lush. Both enjoy buoyant occupancies because of their proximity to a major urban market and easy accessibility from a major international airport. The Sentosa resorts cannot, however, compete with the Whitsundays or Mamanucas properties in terms of their natural setting. They are located on a stretch of water frequented by a veritable armada of container ships and tankers. When the author visited the resort, it was evident that these very visible examples of industrial activity were a discouragement to potential swimmers. The Shangri-La *is* competition for island beach resorts, but the adjoining environment will never be pristine or natural enough to allow it to compete directly in markets which are seeking a natural experience: 'Singapore lacks the water quality, particularly in terms of coral reefs and fish life and the idyllic setting of other regional resorts' (See Tho 1994: 8). Since the technology now exists to create entirely artificial, undercover environments that mimic actual resorts, it may be that sun, sea and sand resorts as traditionally conceived are *only* able to compete by emphasising the attractiveness of their natural and cultural contexts.

Since it is the quality of the natural environment which has constituted the main attraction of the Whitsunday and Mamanuca groups, it would be inadvisable for marketeers to overemphasise facilities as a key attraction. The closest that the Whitsundays have been to a 'facilities fetishism' is the philosophy of Keith Williams, developer of Hamilton Island Resort (Williams 1985). Being the very biggest was an important dimension of the Williams approach and he emphasised aspects such as the fact that the high-rise Hamilton Towers offered the largest hotel rooms in the Pacific, though the resort has also emphasised its association with the Great Barrier Reef. Hayman Island Resort has also emphasised its setting close to the Great Barrier Reef. The tropical lushness of its landscaping and striking architecture do, however, seem out of context when surrounded by the dry barrenness of the island. Having originally sought to create an ambience of 'European elegance' following the redevelopment of the resort, the recent move is back to a more Australian concept, perhaps signalling that 'authenticity' is emerging as a marketing tool. The General Manager of Hayman (T. Klein pers.comm.) referred to the prominence of Australian themes in and around the resort. Finally, the refurbishment of Daydream Island Resort involved a virtual rebuilding which has transformed the tiny island through development on a vast scale. It is indeed an agreeable environment, albeit a very manicured one. Nowhere in the Mamanucas is the focus on resort buildings as prominent as in these three examples.

Advertisements are worthy of detailed study in the present research because they constitute the 'language of consumption' and, as Sack (1992) has claimed, language is the most consistent manifestation of thought. Not all advertisements are believed, of course, and there is evidence that consumers are increasingly sceptical of their claims. Despite such scepticism, Sack has stated that it is less important than achieving consumer acknowledgement that advertising 'be the principal public language of commodities' (1992: 107).

Advertising and related promotional activity often use rhetorical language in which statements are completed or resolved only if the consumer proceeds to make a purchase. This process can be likened to the urge of music (and listeners) to move to a cadence or point of resolution and has become a discourse with which consumers are familiar. Such advertisements offer the prospect of a world without constraints and responsibility, with prosperity and ease within our grasp. The negative dimension (never hinted at in the advertisements themselves) is that unrestricted freedom can also create a weightless and disorientating world. It can be argued that resort guests who have been imbued with such expectations prior to their departure from home are not in a suitable frame of mind to engage in balanced cross-cultural communication.

According to Sack, advertising has a cumulative effect of making the receiver uncertain over the distinction between the subjective and the objective. The uniting of subjective feelings and objective things is achieved through interweaving and interpenetration of the various elements. This trend has intensified as increasing amounts of promotion have been applied to intangible, experience-based 'products' rather than to tangible commodities. The tendency for almost all aspects of life, from religion to politics and sex, to become the subject of marketing has created a 'culture of consumption'. A resort holiday is an experience-based product. Promotion of the components of that product or experience creates a merging of the actual with consumer expectations and interpretations, thereby inventing a new context. Products and experiences brought to our attention through advertising and promotion also bring together the public and the private domains. Products can be attractive because they appear to possess apparently objective and publicly-recognised attributes, but also because they encourage us to impart personal meanings and values. Many of these are prompted by advertising, notably the message that making a particular purchase will help us to resolve any problems that we are experiencing. Promotion has a tendency to draw such problems to our attention and to magnify their importance. Advertising preoccupies us with our own personal concerns and places them within a particular social context by suggesting interactions with other people as one means of bringing about resolution. Urry says that advertising promises to help us 'define ourselves through appeals to personal desires such as sex, power, prestige and/or money' (1990: 119). Advertisements for resorts reassure us about our relations with others through suggesting social situations where we are likely to excel and/or feel comfortable. At the same time they present images which appear to place us in unusual contexts, thereby setting us apart from other consumers and helping us to 'become our true selves' (1992: 120).

Advertising landscapes has the effect of placing them under human control by allowing them to be viewed and experienced. The packaging of nature diminishes its scope since the process of transformation means that it is no longer untamed and untouched – it is 'cooked' rather than 'raw'. Ironically, such landscapes become attractive precisely because the advertisements make them appear untouched and untamed. The 'taming of the wild' is a recurring theme in resort promotions. The fact that lavishly equipped resorts are shown to be located on the perimeter of wild landscapes (such as the Great Barrier Reef) seems to process and refine the natural environment for human enjoyment and recreation. This issue will be taken up in greater detail in chapter 8.

For Debord, resorts are a type of spectacle and advertisements transform us from 'being, to having, to appearing'. Photographic representations transport us to situations where we recognise a number of familiar props and/or images. The settings are typically exotic places consisting of reconstructed local culture and processed natural environment. Advertising also points to the empowering capacity of commodities. Purchasing the right resort holiday can be an entrée into a desired or sought-after social situation. Because most consumers feel insecure about living in a world of strangers, advertisements generally adopt a conservative approach to human interactions which avoid any reference to consumers involved in awkward social situations (except in instances where the product is seen as the solution to a familiar, awkward situation). Nature on the other hand is depicted with what Sack calls 'extraordinary freedom' (1992: 129), typically using emotive language to give it appeal. This is also the view of the Marxist Lefebvre, who has described advertising as the 'poetry of the modern world' offering 'freedom from fear' and addressing the 'terror of alienation' (Lefebvre 1984: xv).

Resorts are part of a trend towards the domination of an ever-increasing proportion of human experience by appeals to consume. Sack (1992) sees the proliferation of 'consumption places' as a symptom of this modernity. Such images offer us the prospect of creating our own contexts in a world which seems to revolve around us. A recent advertisement for the Shangri-La group of hotels and resorts (including reference to their two Fiji properties) promised to 'make your dreams become reality'. As Sack has observed, 'selling the idea of losing oneself to discover oneself is common advertising practice' (1992: 240). Resort advertisements typically feature glamour models and idealise the functioning and life of the resort. It is not surprising that resorts feature prominently in lifestyle magazines as settings for romance and other desirable emotions and activities.

Critics of advertising point to a variety of issues which they claim encourage consumers to be irresponsible, even irrational. Examples from resort promotions include the failure to provide guests with a sense of context for the destination, which results in a lack of awareness of the consequences of their own activities. Others are the absence of an educational focus and the fact that the 'paradise' depicted is typically empty and disorienting. Sack (1992) has argued that promotions ignore context and locate resorts and equivalent consumption places 'out of history and geography' and 'promote irresponsibility which is immoral'. It is of course questionable whether promotional activity should take responsibility for making consumers aware of the repercussions of their own actions. Perhaps that discipline is best shown to visitors at the destination itself, or perhaps public tourism authorities could play a lead role through monitoring destination marketing. Historically, the Fiji Government has placed notices about appropriate behaviour for visitors on the back of disembarkation cards and in *Fiji Magic*, the major tourism magazine. The government is probably best placed to influence advertising messages in campaigns where the private sector seeks dollar-for-dollar funding as part of a joint campaign. Unfortunately, it is often the FVB which is attempting to secure private-sector funding, leaving the government in a weak negotiating position. The increasing interest being shown in ecotourism by the Departments of Tourism in Australia and Fiji may indicate a trend to bring about a stronger sense of public responsibility, though equivalent government tourism promotion bodies are shifting to a more commercially-driven approach in which consumer awareness issues are a low priority. Another solution might be for tour operators to implement codes of conduct which incorporate advice to clients urging the need for responsible behaviour. Currently such activity in Australia is predominantly confined to the so-called 'ecotourism operators' (Richardson 1993) but in Germany, where the environmental movement is very strong, one of the two major operators, Turistik Union International (TUI), already imposes tight controls on ethical standards (involving the natural and social environments) by the destination operators.

According to Sack, the dominant influence played by promotion over consumption places is leading consumers away from reality and into a fantasy world. He has stated that 'when places of consumption themselves evoke the feeling that they are unreal, magical, impermanent and unauthentic, then the very grounds for experiencing reality are shaken' (1992: 4). Our dominant responses to such places are described as normative, arousing emotions such as excitement, a sense of abrupt change and of magic. These are often accompanied by disorientation because we suspect that the resort is shallow and unauthentic, but cannot articulate why this is the case (Sandford and Law 1967).

Urry's *The Tourist Gaze* (1990) is premised on the emerging dominance of the visual over other senses in the practice of consumption-oriented activities such as tourism. Drawing on the work of Sontag (1979), he points to the dominant role of photography in intensifying our reliance on signs and symbols to represent particular tourism sites. He also points to the prevalence of appearance over substance. Photography is of course the key component in resort and in tourism promotions generally. Its influence is particularly powerful since consumers are less cynical of photography than of words, often holding views best summed up in the expression that 'the camera never lies'. The fashionability of the tanned body is an example of the 'embodiment' of this shallowness of experience, in the sense that it is never more than skin deep! Urry (1990) also points to the blending of spectacle and of culture as characteristic features of tourism.

According to Urry, 'the tourist gaze' is 'directed to features of the landscape and townscape which separate them off from everyday experience' (1990: 3). It is the sense of 'otherness' that provides a site with status and value in tourism terms. In expressing why he believes that tourism is a postmodern phenomenon, Urry stressed the integral role of the visual, the aesthetic and the popular, a view shared by Sack. A scene or view is framed, labelled as 'different' and then experienced again and again through photographs or postcards. 'Photographs organise our anticipation or daydreaming about travel' (Urry 1990: 140). Daydreaming is another commonly recurring theme in the sociological literature. Modern consumption is seen as favouring daydreaming and experience aimed at satisfying fantasy. Since the actual experience is destined never to be as good as the idealised version, consumers constantly hanker for more.

Advertising relies on the use of particular cultural reference points and signs to enable potential consumers to absorb the intended message. Writing on the semiotics of tourism, Baudrillard (1983) has proposed that 'the use of the term "fun" in advertising a Club Med holiday is a metaphor for sex. Other signs such as lovers in Paris function "metonymically"' (quoted in Urry 1990: 140). Baudrillard depicts postmodernism as 'an age of simulation, a play of signs . . . what we increasingly consume are signs and representations' (1983: 15). According to Eco (1986), 'the sign aims to be the thing, to abolish the distinction of the reference' and 'things must equal reality even [if] reality was fantasy' (1986: 7, 15). It is interesting to speculate on the extent to which the Mamanucas and the Whitsundays are sold as fantasy products and whether this is the most appropriate approach. Providing holidaymakers with fantasy experiences involves major investments in purpose-built facilities (the various Disney theme parks exemplify the 'high tech' approach) and could face opposition in both island groups. In the Whitsundays the GBRMPA expects developments within the Park to be compatible with the adjoining reef environment. In Fiji, the views of the indigenous population would need to be accounted for (Euro-Disney faced many development hurdles because the resident French population challenged the suitability of the fantasy-based theme park). The current emphasis on the promotion of Fijian culture as an integral part of tourism developments would increase the pressure for fantasy products to be perceived as 'authentic'.

Turner and Ash (1975) identified the close association that promotion attempts to make between tans and good health. Sunbathing is also depicted as a means by which holidaymakers can commune with nature. Ironically, as the impulse to commune with nature appears to be strengthening, the fashionability of sunbathing (though not yet tans) is fading. Nevertheless the ritual of sunbathing is still potent. By increasing the body's visibility, its desirability is enhanced. As Ash (1974: 279) has stated, 'the tan is in every sense a cult with its own obsessive ritual, its sacrifices, its specialised costume and implements'. Like much advertising it is obsessively concerned with health and with youth. The ritual dimension of island resorts is reminiscent of Mumford's description of activity by travellers from ancient Greece on the great health resort and sanatorium, the island of Cos (Mumford 1961: 160).

The imagery of paradise used in resort advertising has social and environmental repercussions. Sack (1992) has described the three essential dimensions of paradise in its original form as (a) a place, (b) where the inhabitants are sentient beings capable of experiencing pleasure and (c) a perfect and harmonious integration of the elements of 'other realms' (principally the realms of social relations, nature and the realm of 'meaning'). The tropical resort paradise differs from longer-established religious concepts of paradise (though one Queensland island resort is solely designated for clients adhering to the Buddhist faith). The traditional type generally demands that a candidate has led a good, moral and purposeful life, often involving some pain. In the secular resort paradise, money is the only entry criterion. In return for a financial outlay, one is promised a pleasant, harmonious place in which each consumer is the centre of immediate gratification. It is a 'cornucopia of food and drink and an environment which gratifies every sense . . . a world of prosperity and ease' (1992: 198). The promotional rhetoric implies that the earlier barriers to a life of indulgence (or if it is preferred, indolence) have been overcome.

In the case of island resorts the product is a blend of facilities and setting. This contrasts with the promotion of tangible commodities, which are often depicted as being used in an unusual context, that can nevertheless appear believable. For example the images currently advertising Fosters beer in Australia consist of a variety of outback scenes where the beer is being consumed, accompanied by a caption proposing that life 'does not get any better than this'. In the case of resorts, the location and hence the setting is fixed, though of course photographic and other representations can substantially transform the perspective from which their surroundings are viewed. Since unusual contexts are suggested so frequently in advertising, there is a great temptation to use such devices in resort advertising, although the true context (the setting in a natural and cultural environment) is there and irrefutable.

Urry has described tourism as 'concerned with spectacle and with cultural practices which partly implode into one another' (1990: 86). The result, in the case of Fiji, is the blending of 'pure entertainment' (appealing to Australian tastes) with elements of Fijian performance and cultural expression. It is interesting to speculate whether cultural performance and spectacle in the Whitsundays could reflect a particular regional context. The current nature of most entertainment on offer is very eclectic and is generally lacking in any sense of regional reference. One difference between the two settings is the origins of the market – predominantly domestic in the Whitsundays and predominantly international in the Mamanucas. There is a conscious attempt to 'showcase' Fijian culture in the Mamanucas through mekes, kava drinking and other cultural activities to an international audience. In the Whitsundays, entertainment and performance may coincidentally be representative of Australian culture but there is little deliberate attempt to make it so, nor to promote it as such.

Resorts are an integral part of a complex production and consumption chain, though the only actual production which takes place inside resorts is of a theatrical, performance type. Whether guests feel a part of the production and consumption network depends substantially on decisions made by individual resorts about their style of operation. Until 1992 Fiji had no television, thereby severing one link with the 'global village'. Though television broadcasting has been instigated since then, many resorts have chosen not to provide television sets in the guest rooms. In the Whitsundays all resorts (including Club Med) except Palm Bay Hideaway and Hook Island Wilderness Lodge do incorporate television sets in each guest room or unit. Their presence is significant because such broadcasting creates contexts for a mass market through image projection. The technology has the power to transport consumers from their specific locality. At the same time it localises the image by bringing them inside the resort unit or room. Though the impact of a single medium such as television should not be overestimated, it is evidence of the all-pervasive influence of the visual over tourism. A 'village-style' resort setting which spurns the presence of television has certainly more scope for conjuring up a 'traditional' community, however contrived it may be.

One frequent message of advertising is that commodities create contexts. An example of this is the print advertisements used by the Fijian Resort on Yanuca Island which describes itself in the terms 'the Fijian is Fiji'. Urry also describes holiday centres (and one might include island resorts in this category) as examples of the 'aestheticisation of consumption' (1990: 14). Others have argued that this is a secular reinterpretation of a previously religious motif. Leers (1983: 3–38) has stated that advertising (unintentionally) reinforces and spreads a culture of consumption in which the presence of material goods has replaced the pursuit of religious goods. Resorts could be thought of as providing a type of 'instant paradise' for those who no longer have faith in an afterlife. For Boorstin (1973) consumption is the principal device that binds our culture together. This role is certainly evident to the extent that travelling to an island resort enables us to express personal and group identities in a world of strangers, albeit a curious blend which sets individuals apart, and then draws them together.

Images and the responses that they elicit have the power to transform the meaning of resort activities. Urry's statement that 'the visual gaze renders extraordinary, activities that would otherwise be mundane' (1990: 12) gives a wider context to Tuai's assertion (pers.comm.) at Beachcomber Island that guests have shown an increasing interest in spectating when staff undertake rethatching of the bures. What may have appeared a mundane activity to the perpetrators, was transformed into a spectacle. The power of imagery can transform such events into 'must do's' in the sense that such 'events' are seen as capturing the 'essence' of the destination. The transformation of the ordinary into the extraordinary can also be applied to activities such as swimming or walking which would be considered unexceptional by the exponents in other circumstances. Such activities become 'special' because they take place in a particular island resort setting. The influence of such settings over activities is exemplified by a series of Club Med print advertisements reproduced and critiqued in Thurot-Gambier (1981). These advertisements, in the form of colour sketches, are each accompanied by a brief caption using words such as *Rire* (laugh), *Jouer* (play) and *Manger* (eat). The implication of such advertisements is that fundamental human activities and emotions are best experienced in the types of setting offered by Club Med Holidays.

In selecting a holiday, most consumers appear to seek a blend of reassurance and novelty. Plog's (1974) continuum linking extremes of allocentric and psychocentric behaviour has suggested that intending travellers can be placed at different locations along the axis. Resorts appeal to certain personalities because of the contrast that they offer with daily life and with the home environment. Others seek a greater amount of reassurance and familiarity. Most probably seek elements of both. What do we know about the mental processing of imagery on resorts and other tourism places? Levi-Strauss (1970) has depicted the human mind as being filled with some opposing and some mediating concepts and that these are used to impose a kind of mental order on the natural and social worlds. The clash of the opposing concepts creates ambivalence in the mind of consumers when confronted by certain propositions. One example of such ambivalence was evident during the focus discussion groups undertaken for the present research. When asked to respond to a variety of prompts, participants expressed a liking for the word 'island' (it seemed natural), liked the word 'resort' less (might be overdeveloped, but offers the reassurance of modern conveniences), liked the use of local sounding island names (for example, 'Mamanucas' sounds authentic) as opposed to tourist names (for example, 'Castaway Island'), but were reassured by the familiarity of the latter. The responses were indicative of a marked ambivalence to the 'raw and the cooked' dimensions of resorts and their settings.

Applying related concepts to tourism, Bachelard (1957) has argued that the idea of an island resort triggers a dialectic of theses and antitheses in the mind of the consumer. The opposing concepts include ideas of 'the inside and the outside, the latent and the manifest, the closed and the open' and the 'introverted and the extroverted'. Islands are reassuring, he has argued, because of their confined space (experiencing the 'inside') and the 'outside' (somewhere different and new in a vast ocean). He describes the island experience as 'less familiar than the home or suburb'. The desire for contrast with the home environment was evident in certain of the focus discussion groups. Participants who had travelled with young children commented on the *insecurity* of the suburbs of large cities, especially for children, and the fact that the Mamanuca islands offer the prospect of a secure environment – at least in relative terms. The confined space of an island where all corners can be reached by walking (in the case of most of the Mamanuca and Whitsunday resort islands) contrasts with even a suburban trip to a milk-bar where a car may be essential and where parents may need to supervise children closely to avoid traffic. This conforms to the view expressed by Popcorn (1991) that consumers are tending to turn inwards and are searching for security and refuge in an increasingly violent and transient world. She describes this process as 'cocooning'. Resorts may be able to satisfy some of these security concerns. Ritchie (1993) has argued that those destinations which prove themselves best able to provide such reassurance will be the most successful operations. It seems likely that some consumers will respond well to propositions or messages promising the prospect of total contrast, whereas in other cases a more complex blend of familiarity and contrast will be preferred. Extreme care is needed in achieving a balance appropriate to the resort's chosen target markets.

Various authors have examined the perceptions held by consumers and by the travel industry in the developed world of developing country destinations (Silver 1993; Britton 1979). Thurot-Gambier (1981) categorised such perceptions under six headings, the first three of which might be described as 'gratifications' and the latter three as typical distortions. The six headings are: familiarisation, reassurance, excitement, the absence of 'production', social life reduced to folklore and servility confused with hospitality. Resorts sometimes draw parallels with better-known destinations as a means of conveying quality and status. The present author's own investigations identified a brochure on Bowen and the Whitsundays estimated by the author to date from the 1930s, which included the comment 'Naples or Rio de Janeiro minus its bay and Sydney or Bowen without its harbour would be inconceivable' (QGTB undated). In both the Whitsundays and the Mamanucas, the balance between the exotic and the familiar is an uneasy one. Traditionally, Fiji has conveyed a strong flavour of the exotic, but often accompanied by the reassuring message that many other Australians go there too. The Whitsundays have probably been *disadvantaged* as tropical islands because as a domestic destination they convey less of the exotic. In both cases much of the reassurance comes from the use of the word 'resort' and the promise of modern, quality facilities. These messages offer a predominantly contemporary flavour (in the case of the Whitsundays) and a more traditional flavour (in the case of the Mamanucas). The preference of consumers in various discussion groups for the 'exotic' island names (in the Mamanucas those with traditional Fijian-sounding names), but at the same time attraction for English terms such as Plantation, Castaway, Beachcomber and Treasure and reassurance by the word 'resort' is an indication that generalisations about the level of familiarity for domestic and international destinations are risky. An additional dimension is that the level of reassurance demanded by holidaymakers may be changing. Inexperienced travellers often prefer to be surrounded by people of their own nationality to minimise the apparent strangeness of an overseas destination. More experienced travellers may welcome the contrast with home provided by a cultural and ethnic mix. An apparent preference emerged amongst Australian travellers in the focus discussion groups for those island resort destinations, international or domestic, which drew their clientele from a wide range of countries.

TROPICAL ISLANDS AS THE ULTIMATE 'GRATIFICATION'

Sociological analyses of the imagery disseminated by media advertising provide insights into the nature of island resort promotions, though the heavy concentration on west European and American consumers needs to be acknowledged as a limitation. The extent to which they can be applied to the emerging economies of the Asia–Pacific region merits further research.

In the following section the type of promotional image projected of resort areas such as the Whitsundays and the Mamanucas is evaluated in terms of fulfilling needs and (more particularly) desires. This is done by applying the twenty-four 'gratifications' that, in Berger's view (Berger 1991), are the foundations of the language of advertising. To what extent do resorts, and those who promote them, set out to offer such gratifications? It is suggested that understanding resort promotions in terms of the

Table 6.1 Berger's gratifications: those applicable to the Whitsundays and the Mamanucas

Gratification concept	Application to resort promotion
1 To experience the beautiful	Island resorts are frequently promoted as beautiful places.
2 To share experiences	Promotions emphasising 'togetherness'. Very few resort tourists travel alone (resort managers pers.comm.).
3 To satisfy curiosity/ be informed	The desire to see and experience a new and desirable place.
4 To identify with the Deity and the Divine plan	Tourism as a 'secular pilgrimage' to an 'earthly paradise' with overtones of the Garden of Eden.
5 To find distraction and diversion	The heavy promotional emphasis on the diversity of resort activities.
6 To experience empathy	Resorts as places which can help families and other groups rediscover relationships in a relaxed environment.
7 To experience extreme emotions in a guilt-free and controlled situation	The emphasis on sex free from home constraints. The redefined Garden of Eden in which woman can err without fear of retribution (Wernick 1991). Emotions such as laughter, play, a sense of awe and excitement (cf. Club Med advertisements).
8 To identify models to imitate	Promotions which refer to famous personalities (e.g. Hamilton Island and ex-Beatle George Harrison) and as a playground for the 'rich and beautiful'.
9 To gain an identity	The Brampton Island promotional theme 'Discover Yourself'. The theme of self-fulfilment in advertising.
10 To gain information about the world	This aspect is not pronounced, but the idea of a simpler, village-based lifestyle offers the prospect of new insights into the world.
11 To reinforce our belief in justice	The 'earthly paradise' as a place devoid of conflict which implies justice (though it is a world accessible only to those with money).
12 To believe in romantic love	The promotion of honeymoon packages using classic 'romantic' settings (e.g. sunset on a palm-fringed beach). The staging of weddings (Hamilton Island has its own chapel).
13 To believe in magic, the marvellous and the miraculous	The Great Barrier Reef as a 'Wonder of the World' and the seventy-four Whitsunday islands as a 'marvel of nature'. Fijian mythology as conveying 'atmosphere'.
14 To see order imposed on the world	The consumer as 'centre of the world'. The compactness of the island and exclusivity offer order and security.

Table 6.1 cont.

Gratification concept	Application to resort promotion
15 To participate in history	The appeal to a nostalgia for a simpler village life (exemplified by the *bure* accommodation in the Mamanucas).
16 To be purged of unpleasant emotions	The application of suntan oil as a balm, purifying the recipient and evoking fantasy.
17 To obtain outlets for our sexual drives in a guilt-free environment	The 'worship' of the body beautiful through tanning and exposure. The previous Great Keppell Island promotion 'It's a Great Place to Get Wrecked' depicting the ease of finding sexual partners.
18 To be amused	Humour is uncommon in tourism promotion (Thurot-Gambier 1981), though she cited the Club Med advertisement entitled *Rire* (laughter). Photos of groups of youths laughing together usually in physical water-based encounters are common.
19 To affirm moral, spiritual and cultural values	In entering the popular consciousness the island paradise myth reinforces the sociocultural desirability of an island resort holiday, offers a contrast with everyday materialism and a guilt-free experience.

Source: Berger 1991 (adaption to resorts by the author)

gratifications may assist resort and destination managers in taking decisions about what line of persuasion to use in resort promotions. The discussion then delves deeper than the promise of instant gratification by asking whether the type of appeal is a creation of modern consumerism.

Tables 6.1 and 6.2 highlight the twenty-four gratifications, separating them into those (nineteen) which appear to apply directly to the examples studied in the Mamanucas and the Whitsundays and those (five) which do not. The predominance of the former is a striking feature.

A number of Berger's 'gratifications' are less immediately applicable to island resort promotions, though this is not to discount the possibility that imaginative marketeers could successfully draw upon them. These are shown in table 6.2.

It is remarkable that resorts appear to lend themselves to nearly all of the gratifications noted. The fact that resorts offer a 'slice of life' over an extended period, such as a week or a fortnight, appears to draw upon most of the gratification messages used in promoting a wide range of products and incorporating many of the desires embodied in the consumer culture.

Table 6.2 Berger's gratifications: those not applicable to the Whitsundays and the Mamanucas

Gratification concept	Why not applicable
To see authority figures exulted or deflated	Competitive sport and other activities involve certain individuals 'winning' and others 'losing' with role-playing, inversion of roles and ritualistic exultation or humiliation. The 'ludic' is certainly a significant aspect of resorts (Huizinga 1950).
To see others make mistakes	Again this can play a part in the ludic or playful aspects of resort activities, but is not an obvious part of resort promotion.
To explore taboo subjects with impunity and without risk	Because resorts offer relative anonymity (e.g. the use of first names in preference to surnames) they might be seen as offering a shelter for certain taboo activities.
To experience the ugly	This is one area that resorts do not promote. The primary emphasis is on beauty and ironic humour is rarely displayed.
To see villains in action	Again the 'performance' dimension of a resort means that anything is possible, though usually on a symbolic or acted-out basis. Gottlieb's view of 'King or Queen for a Day' role-playing may apply.

Source: Berger 1991 (adaption to resorts by the author)

WHAT THE HOLIDAYMAKERS THINK

This chapter has focused on interpretations by sociologists and related social scientists to issues of tourism advertising and imagery. In the final section, some key findings of the focus discussion groups are summarised to provide an Australian consumer perspective which can in turn be related to some of the theoretical principles considered earlier and then applied specifically to the Mamanucas and to the Whitsundays.

Consumer images of resorts were predominantly positive, with tropical islands seen as embodying the most gratifying associations of the word 'resort'. Some anxiety was expressed about the fear of being trapped, though this was less prominent than the positive attractions. Participants were aware of individual islands in Queensland, but were generally only aware of overseas destinations in terms of brand names such as Club Med, or whole countries. There was a fairly strong awareness of location in the case of the Whitsundays and virtually no awareness of the name Mamanucas at all.

The Whitsundays were thought of as up-market and luxurious, with little awareness of the two smaller and lower-key properties. In the case of the Mamanucas, indigenous-sounding names were perceived as exotic, whilst the word 'resort' provided reassurance of western-type facilities. The word 'island' conveyed a sense of romance better than the word 'resort' and appeared to fit the concept of the 'tropical island myth' as depicted by sociologists. The key concepts associated with the Whitsundays were relaxation, sporting opportunities, hedonism, glamour/prestige and beautiful scenery. The main negatives were a perception of expense, overdevelopment in certain places and questionable service. The Mamanucas were seen as offering the prospect of great diversity, were suited to children, less expensive, romantic and underdeveloped.

Drawing firm conclusions from the deliberations of the focus discussion groups is of necessity tentative because of the limitations of the methodology already referred to. Nevertheless a number of potentially valuable findings can be observed. The Whitsunday and Mamanucas examples suggest that Australians have a better knowledge of areas within their own country than of equivalent areas overseas. The familiarity of domestic destinations is simultaneously a strength and a weakness for resort and destination marketing. Familiarity provides consumers with reassurance about issues such as infrastructure, health and other services, language and currency to name but a few. The unfamiliarity of overseas destinations, however, is associated with the exotic, with 'otherness', with cultural diversity and with historical interest. Whilst the myth of the idyllic tropical island applies to some extent to both areas, it has greater power over the consumer imagination when applied to overseas South Pacific settings such as Fiji. This gives these destinations a marketing advantage.

Familiarity also seems to lead to an impression of proximity – the Whitsundays are perceived as much more easily accessible than the Mamanucas, though in reality there is little difference. This view of accessibility gives a natural advantage to the domestic destination in the short break (3–5 night stay) market. Heavy marketing of such packages has benefited Queensland relative to international destinations such as Fiji in this growing market area. The larger marketing budgets available to domestic promoters have assisted this trend.

One disadvantage faced by the domestic destination and one based largely on a misconception, is the view that domestic holidays cost more than equivalent international holidays. The perception is based on the previously high cost and highly regulated domestic airline environment and does not seem to have been greatly influenced, as yet, by the advent of domestic airline deregulation in 1991. There is evidence that attitudes are changing. Participants were aware of an 'increasing number of special packages' to the Whitsundays. Whilst domestic means less exotic in the minds of consumers, the idea of the Whitsundays as a highly desirable destination is growing. This is possibly symptomatic of an increasing interest in domestic holidays by Australian travellers.

This chapter has highlighted the minimal importance that Australians place on tour operators and/or tour wholesalers for both domestic and international holidays. This contrasts with the state of affairs in most

European generating countries where tour operators play a pivotal role. The study reveals some insights into social attitudes ('Australians are more friendly when they are travelling overseas'). The preference by a number of participants for Australian resorts which attract an international clientele is an indication of the greater integration of Australia into world tourism, with the previously tiny number of overseas visitors growing rapidly and providing many Australian destinations with a more cosmopolitan atmosphere.

The study also highlights the danger of overdevelopment in areas of great natural beauty, even where it is geographically confined. The influence of one highly developed resort (Hamilton Island) has coloured overall consumer impressions of an area which is otherwise notable for its world heritage qualities. Consumers perceive the smaller Fijian bure-style resorts as more compatible with the environment, despite the lesser stringency of environmental regulations in Fiji. The whole issue of perceived and actual environmental standards is likely to play an increasing role in consumer choices between domestic and international destinations. The fact that the image of the Whitsundays was dominated by high cost and luxurious resorts (a result of the large marketing budgets of specific exclusive resorts such as Hayman and Hamilton) points to a mismatch between image and reality, since most resorts, in fact, cater to middle-income holidaymakers. The resulting misconception is indicative of the dangers of individual resorts undertaking major advertising initiatives independent of the regions in which they are located. The need for greater collective marketing to bring about a more consistent image is evident. Despite this, caution should be observed over the choice of image to represent the destination. Queensland's current tourism logo is a palm tree, consistent with the stereotypical tropical island image. A number of respondents commented on the paucity of palm trees in the Whitsundays. In reality those that do exist were introduced; the typical vegetation is dominated by eucalypts! Images should not be based on stereotypes if consumer disappointment ensues.

This chapter has shown the endurance of the escapist imagery of the 'tropical island paradise'. The image was formed in the eighteenth and nineteenth centuries but has been reinforced by the increasing democratisation of travel and by the dissemination of promotional material to an ever larger audience. The commercialisation of cultural images and the extent to which these materialise into attainable consumer experiences has enabled a phenomenon which was already a potent force prior to the twentieth century, to become a seemingly universally sought-after consumerist ideal – to while away one's days on a desert island. Commercialisation has of course transformed the 'desert island' into a place frequented by hedonists in search of sun, sea and sand, inhabiting condos, hotels and other manifestations of mass tourism. This is not to say that all tropical island resorts are equally commercialised. The Mamanuca resorts are still relatively small scale and, though 'commercial', do fulfil the tropical island image through the use of self-contained bures on the beach. In comparison, this style of accommodation is a tiny minority in the Whitsundays where the major hotel chains have institutionalised the holiday experience, sometimes in relatively subtle ways. The responses from consumers and from travel agents is suggestive that in its most commercial form – Hamilton Island – the Whitsundays may have moved too quickly and too far from the original tropical island image.

The tropical island image is not entirely a blessing. The South Pacific islands have encountered the frustration implied in the expression – 'this is more than a beach, it's a country'. The image has a potent drawing power, but there are many countries that can offer the 'raw materials' that are regarded as delivering that image – a tropical climate and a beach. A challenge for both the Mamanucas and the Whitsundays is to recognise the existence of the image, to build and expand upon it, rather than simply delivering sun and sand in an ever more enticing manner.

7

SOCIAL ENGINEERING?

Are island resorts really 'paradise on earth' as they are so often portrayed in promotional messages, particularly those depicting the South Pacific? Or are they the very antithesis of their promotion? The British writer David Lodge in his novel *Paradise News* (1991: 143) refers to one Pacific destination, Hawaii, as 'Paradise lost . . . Paradise stolen, Paradise raped, Paradise infected, Paradise owned, packaged, Paradise sold'. Chapter 6 included a discussion of the evolution of images of tropical islands over a period of two centuries. This chapter evaluates the type of social relations which occur – a paradise does after all necessitate the presence of sentient beings. Can island resorts be the settings for meaningful social interactions? Since most resort islands were uninhabited prior to tourism and the tenure of resort guests is short, can an island resort be a 'community'? Urry (1990) justifies his depiction of tourism as a key example of postmodernism by citing examples such as the prevalence of pastiche, a preoccupation with surfaces and image, the use of landscape as theme and an apparent hostility to tradition. If it is found that such features are common to island resorts, then we may regard them as 'postmodern communities'. Are island resorts a type of 'tourism society' or 'short-term society' where people gather in a 'group culture' displaying 'appropriate behaviour' living at close quarters and interacting on a fairly intimate, if superficial basis (Foster 1986)? Wagner (1977) has proposed that we are best able to understand the interactions between visitors and local residents by conceptualising resorts as special types of community (she uses the term *communitas* in preference to community). Treating interactions purely from the perspective of cultural difference is inadequate in her view and needs to be supplemented by a thorough appraisal of resorts as communities.

The idealisation of island resorts has been analysed by sociologists and social anthropologists. Are resorts characterised, as such critics claim, by a predominance of the balance sheet (Stettner 1993), by the absence of democracy, by voyeurism and intrusiveness, by hyperreality and living in the space of absolute simulation (Eco 1986), by coerced consumption (Cohen 1982), by ritual and fetishism (for example, suntans) (Ash 1974), by an absence of history and intolerance of minorities (Britton 1979)? If, as some critics imply, resorts are founded on a concept of vacuousness, the very attempt to involve them in sustainable tourism development may be doomed to failure. Or do we regard the resort experience as an example of spontaneous *communitas* offering consumers a useful contrast with their normal lives?

In examining the critiques by sociologists of 'imagery' and its impact on social relations, we should be conscious of a frequent deficiency in the approach. Critics often fail to distinguish between the terms 'imagery' and 'advertising'. As previously stated, 'induced imagery' or marketing constitutes only a small element of the overall image-building process. Within marketing, the promotional mix in turn is only one of a number of elements. Finally advertising is only one of four sub-elements within the promotional mix. Advertising, then, is a fairly small, albeit conspicuous element of marketing. It is true that advertising agencies assist their clients by developing and placing messages in a variety of media and their influence extends into other aspects of the promotional mix including public relations and sales promotions. Nevertheless, the assumption made by some sociologists that advertising exercises a great influence over human behaviour is questionable.

The social dimension of travel to island resorts is many-faceted. It includes peer group pressure during the decision-making process, interaction with those accompanying the traveller (for example, family or friends), encounters with staff and with the local population on the island, interaction with other guests and the social context of 'holiday tales' on the return home (Thompson 1981). Island resorts include a variety of social contexts worthy of observation, notably the contrasting social environments of urban generating regions (such as Melbourne and Sydney) and of the resorts themselves. Even Hamilton Island is intimate in comparison to the metropolis. If its 700 rooms are occupied by an average of 2.5 persons each, the population of houseguests never exceeds about 1,750 – much smaller than any suburb of Melbourne or Sydney. Social relations within resorts are played out in 'convenience' as opposed to 'obligation' based encounters. In such settings, it is asked, should the resort manager be a kind of town/village mayor, or is he/she simply a hotel manager with certain ancillary responsibilities? Another way to understand island resort life is as a deliberate and necessary contrast with the 'norm-governed institutionalised abstract nature of social structure' typical of everyday life (Turner 1984a: 127). This approach is exemplified in the writings of Wagner (1977), Turner (1984b) and McCracken (1982). Such approaches emphasise the rituals of resort life as being anti-structure, though as ultimately reinforcing of prevailing structures.

This chapter is based on the author's application of broad social theories, many applied to island resorts for the first time. These theories have been supplemented by social insights provided by survey respondents and by field observations made by the author, who sought views and perceptions of the comparative friendliness and sociability of the resorts and the 'ambience' of island resort life. Styles of customer service, including staff and guest interactions, were also evaluated including issues emanating from staff training and recruitment. Do specific Fiji/Australia and/or Mamanuca/ Whitsunday contrasts emerge? And what about the 'routine of the day'? Are resort guests active or passive? Travel agent respondents offered the most objective views, based on their access to a wide range of comparative information on the suitability of island resorts to particular groups of holidaymaker. Consumer and manager impressions were useful in demonstrating the contrasts between the intentions of resort managers and the perceptions and actual experiences of visitors to their resorts.

Tourism researchers have neglected empirical work on the social dimension of island resort life, particularly related to specific localities such as the Mamanucas and the Whitsundays. The major relevant literature consists of social impact studies. These have been undertaken on both Fiji (Plange 1984; King, Pizam and Milman 1993; Pizam, Milman and King 1994); and on Queensland (Cameron McNamara 1987; Pearce and Moscardo 1987; Ross 1990). Social impact studies are often an integral component of major tourism developments (though not often enough according to critics). The recent feasibility study for a proposed Club Med resort at Byron Bay in Northern New South Wales (Plant Location International 1993), for example, includes some useful baseline data for the Club Med development at Lindeman Island in the Whitsundays. Taken collectively, social impact studies have been inconclusive about whether tourism has beneficial or deleterious results. Studies measuring and/or assessing tourism impacts on a pre-existing community are of limited use here since all but two resort islands were uninhabited prior to development. This is not the case for the mainland adjoining such island groups, raising questions about the relationship between resorts and their hinterlands. What are the social policy implications across regions where some sub-regions experience high concentrations of visitors and low resident population densities (typically the islands) and others have low visitor densities and high resident populations (typically the mainland)? In considering social aspects we cannot consider resort islands as totally self-contained, but must acknowledge regional linkages.

Pearce (1981) studied the experience of holidaymakers over the duration of their stay on two Queensland island resorts. In *The Social Psychology of Tourist Behaviour* (1982), he used some relevant examples of holiday behaviour at resorts including South Molle Island in the Whitsundays. Craik's study *Resorting to Tourism: Cultural Policies for Tourism Development in Australia* (1991) and a related article by the same author (1988) have taken a broad sociocultural approach to tourism with a particular emphasis on the

Queensland islands. The more business and management oriented literature often has a social dimension, notably Gee's *Resort Development and Management* (1988) which includes an examination of the role of resort staff and management options for achieving guest satisfaction. Such texts are limited by their narrow business operations framework. The leisure sociology literature offers some insights. The term 'commercialisation of leisure' was first used by Dumazedier (1967) and subsequently developed by Parker (1983). When applied to resorts, Dumazedier's theory points to the increasing proportions of our leisure time dominated by activities which have a commercial underpinning. Resorts exhibit two contrasting dimensions. Activities such as bush-walking are not commercial in character, whilst access to free-to-air television involves being plugged into the commercial leisure network.

A number of studies have evaluated consumer perceptions prior to their departure from home, perceptions and interactions during the trip and the subsequent attitude change after their return. Others have evaluated the social significance of international travel experiences on residents of the generating country. *The Tourist Gaze* (Urry 1990), *The Tourist* (MacCannell 1976), *Empty Meeting Grounds: the Tourist Papers* (MacCannell 1992) and *The Social Psychology of Tourist Behaviour* (Pearce 1982) have all evaluated the social significance of outbound travel. Pearce has also written on changes in visitor attitudes before and after the actual holiday. Sack's *Place, Modernity and the Consumer's World* (1992) takes a geographical perspective, drawing upon examples of shopping malls and theme parks as well as resorts. These texts avoid the compartmentalisation of tourism and relate the phenomenon to global social forces.

Some of the anthropological and sociological literature depicts holiday behaviour as a form of ritual and is derived from van Gennep's concept of 'rites of passage' (1960). The anthropologist Victor Turner (1984a, 1984b) has written extensively on tourism as a deliberate and necessary contrast with the highly structured nature of everyday life. Wagner applied Turner's ideas in her study of Swedish package holidaymakers to the Gambia, West Africa. She has depicted their activity at the destination as a form of 'spontaneous *communitas*' and observed the repercussions when visitors made contact with the local population, caught up in their more highly structured daily existence.

ISLAND RESORTS AS THE REALISATION OF AN IDEAL COMMUNITY

Some of the most potent imagery of the Pacific is of 'deserted' tropical islands. Many islands have traded on the imagery of Robinson Crusoe. Crusoe was a solitary on his island until he observed a footprint in the sand. In the last decade, the footprint has become a potent symbol in tourism advertising, perhaps providing a reassurance to potential travellers that company is not far away and points to a clear link between literary works and tourism promotion. Apart from the footprint, the deserted tropical island implies a solitary existence, far from other human beings, the very antithesis of sociability.

Island resort promotions typically depict their guests in social settings and not alone. Brampton Island has targeted the honeymoon market using 'romantic' brochure photographs of couples in tranquil settings to attract this group. Promotions by other resorts depict (most commonly) families or (in the case of resorts targeted at the singles market), groups of cavorting youths, male and female. The promotional depictions of island resorts would lead us to believe that consumers seek a 'social experience' at the destination, even if that social interaction is confined to a single partner. The preference of most holidaymakers for company is confirmed by the academic literature. Ironically, those resorts attempting to attract young singles base their appeal most blatantly on the social dimension, implying that participants may meet the partner of their dreams!

A resort consists of a physical environment, a workforce and a group of consumers brought together for periods of limited duration (say a week to two weeks). In most cases the rationale for operating a resort is to make a profit (though one of the Mamanuca resorts, Navini was developed to make possible a particular lifestyle for the owners), or to earn a specific return on investment. Fundamentally, it is a financial imperative which underpins interactions between the various participants. Though one should not underestimate the potency of word-of-mouth communication, in most cases consumers are made aware of the resorts through paid-for promotional activity undertaken on behalf of the destination. This activity will influence their expectations and perceptions.

The anthropologist Foster (1986) has applied the term 'short-lived society' to a South Pacific cruise. Like cruises, resorts entail guests sharing a confined space over a prolonged period with other holidaymakers. Sack applies the term 'society' to 'people . . . who have lived together long enough to follow without question, the formal and informal rules that govern their interaction' (1992: 218). His comparison of a small special-interest cruise and a mass market, more commercially oriented cruise demonstrated that different types of cruise participants seek differing levels of exposure to local society and culture.

In certain respects, resort guests may be less mobile than cruise passengers who have the opportunity to disembark in various ports. The fixed location of resorts is exacerbated as a management problem by the unlikelihood of finding another use for resort facilities in the event of financial failure (city hotels, in contrast, can be readily converted into office accommodation).

Bernard Smith has argued that trade acted as a mediator between a stone age and a European society during the exploration of the Pacific. The transformation of the pre-colonial image (and reality) of Fijians as practising cannibalism to their contemporary reputation as generous hosts, is indicative of the power of international trade to transform behaviour.

MacCannell (1992) describes community as 'a crucible in which formed human type and a distinctive social conscience . . . . Every human community once had a specific character, identity, a manner of existence in which its residents also partook, accepting as their own its problems and its virtues' (1992: 87). He emphasises the importance of recognising that differences exist between members within communities, simple human responses to the 'social contract', that is, under what conditions and agreements one lives within this particular 'community'. To MacCannell, resorts are not communities but at best 'postmodern communities'. He casts doubt on even this designation by describing such entities as 'some other kind of social formation under the stolen label of community' (1992: 92). Resorts which attempt to recreate 'traditional' settings, according to MacCannell, fail to address the fundamentals of community. This is because they do not allow for the expression of opposition which, he argues, must be a feature of even the most traditional of societies. He is critical of what he describes as 'too much warm talk' about community, a trend he describes as 'based on a desire to retreat into simpler, ego-centred social forms, a retreat into a kind of primitive fiction'. He sees the fact that the resort rationale is based on financial accounting as incompatible with the foundations of true community.

Jung took a less prescriptive view of community. He wrote of a 'collective unconscious' growing up 'spontaneously, at any time, at any time and without any outside influence' (Jung 1959: 13). He stated that an 'archetypal symbol of community' can arise through a combination of individuals. MacCannell criticises the Jungian view on the grounds that the solidarity to which he refers is 'narcissistic' and 'need not, indeed cannot go beyond a mutual monitoring of appearances' (1992: 93). He implies that the dominance of surface and of image over substance prevents the development of a 'community feeling'.

MacCannell's circumscribed definition of community is evident in his criticism of suburbia which he depicts as lacking in 'community': 'how does the suburban community undertake to make itself a better place to live? By ridding itself of peoples who are different. By transforming itself into a highly self-interest motivated simulacrum of the "primitive isolate"; a fantasy based on a fantasy' (1992: 293). To MacCannell the fact that holiday-makers go to resorts to seek escape into fantasy, precludes them from pursuing the more worthy aspects of community life. MacCannell maintains that resort marketing seeks to attract specific social strata and in effect reproduces the type of social exclusion practiced in certain types of suburbia (most Melburnians and Sydneysiders live in suburbs). MacCannell's critical view of suburbia is challenged by a number of more pragmatic interpretations which warn of the dangers of (usually negative) stereotypes (McCalman 1994). It may also be argued that MacCannell's interpretation is based on a particular variety of North American suburb which has no direct parallel in Australia.

Borrowing from the work of Riesman (1950), Sack (1992) has depicted modern life as a 'world of strangers' in which money is the mediator and regulator of social relations and in which people have difficulty finding shared meanings. Greater 'familiarity and trust' is shown to strangers than was the case in the premodern era (1992: 111), but such trust is an outcome of a particular type of marketing which aimed to bring specified types of people to particular resort settings as consumers. According to Sack, consumer promotions promise products that will help to resolve the loneliness that contemporary humans feel because 'reality has no meaning other than what a person chooses to impart to it' (1992: 45–6). Sociologists have keenly debated the issue of whether consumption exemplifies what Marcuse called 'false consciousness' in that consumers allow themselves to be lured by marketing, promotion and manipulation and hence are precluded from the search for true meaning. The debate can also offer insights into the social dynamics of resorts. Anthropologists have generally recognised the possibility of a 'search for meaning', even in unlikely venues, by characterising tourism as a form of secular pilgrimage incorporating rites of passage and other symbolic activities.

Can one conceive of island resort holidaymakers as nascent communities? MacCannell states that 'an aggregate of individuals, even "demographically similar" individuals, is the opposite of a community' (1992: 294). He claims that photographs in UK travel brochures rarely depict minority groups as consumers. Promotions targeted at Australian consumers exhibit similar selectivity, despite the increasing ethnic diversity of the Australian population. Where minorities are depicted, they are more commonly portrayed as 'exotic' local people who will add 'colour' to the holiday. In Turner and Ash's classic study of tourism (1975), the *Golden Hordes* are

> at the centre of a strictly circumscribed world. Surrogate parents (travel agents, couriers and hotel managers) relieve the tourist of responsibility and protect him/her from harsh reality. Their solicitude restricts the tourist to the beach and certain approved objects of the tourist gaze.
>
> (1975: 7–8)

Cohen takes a milder view, depicting holidaymakers as a collection of distinct groups each deriving a different satisfaction from their holiday experience and seeking different intensities of social and cultural interactions, maintaining that individuals experience different levels of awareness and enlightenment. Broadly speaking this is the perspective taken in the present research, which asks whether island resort travellers seek enlightenment or simply indulgence.

Describing contemporary holidaymakers as 'posttourists', Feifer (1985) provides a more positive interpretation of the contemporary tourism experience. To the posttourist, travel is a game and the object is to delight in a multitude of choice:

> now he wants to behold something sacred, now something informative to broaden him, now something beautiful to lift him and make him finer; and now something just different because he's bored . . . . The posttourist is freed from the constraints of high culture on the one hand and the untrammelled pursuit of the pleasure principle.
>
> (Feifer 1985: 269)

Feifer implies that it is no longer appropriate to think in terms of 'packaged' tourism, as Turner and Ash did (1975). Like Urry she argues that consumers are increasingly unwilling to be treated as part of an undifferentiated mass. Tourism is no longer an activity which lends itself to caricature because 'the distinctiveness of the tourist gaze is lost as such gazes become irreducibly part of a postmodern culture' (1985: 269). Tourism, she implies, has become so much of a part of the way that we see the world and our tourism activities reflect the rest of our lives. This willingness of 'posttourists' to experience very diverse and contrasting experiences certainly points to some opportunities for resort managers. It is often assumed that 'mass packaged tourists' will have no interest in environment and culture. The increasing openness to a 'kaleidoscope' of experiences suggests that this may no longer be the case. The 'posttourist' paradigm certainly offers some insights into the type of social interactions that resort guests may find satisfying. The emergence of the 'posttourist' is evidenced by recent market research using the cluster analysis evaluation technique. Cluster analysis suggests that an increasing number of tourists are searching for a kaleidoscope of experiences, for example a blend of culture, sporting activities and relaxation. The results of such research defy some established tourism categorisations exemplifying a more rigid practice of tour packaging.

One reason for taking the resort seriously as a type of postmodern community, is the view that 'having a history' is less important as a necessary precondition than was previously the case. If we were to accept Fukuyama's (highly controversial) view that we are living in a 'world beyond history', resorts could be regarded as key venues in the search for identity in the post-Cold War period. Fukuyama (1989), Deputy Director of the US State Department Policy Planning staff, first used the expression 'end of history' to describe the world after the fall of Communism. Fukuyama refers to 'a truly universal consumer culture that has become both the symbol and an underpinning of the universal homogenous state' (1989: 10) and to a 'powerful nostalgia for a time when history existed'. The materialisation of what had earlier been described as the 'American dream' by Kojeve points to the more liberal view of the resort as a special form of community, embodying features of both modern and postmodern existence. Though MacCannell sees the 'doctrine of historylessness' and the increasing prevalence of historyless places as negatives, resorts should not be assumed to be places of meaningful social interaction, given a world in which 'historylessness' is increasingly the norm. Fukuyama's views of nostalgia provide useful insights into the motives of resort island guests. Pandering to nostalgia may prompt the creation of settings which appear to be devoid of any conflict resulting in an experience of unreality and alienation. On the other hand, the quest for nostalgia may create an opportunity for managers to acknowledge the local historical background of the resort region or locality and present this to their guests. In an increasingly consumer-driven world, the resort may be more symptomatic of the way that consumers wish to see the world than at first appears the case.

MacCannell's apparent resistance to concepts of the resort 'community' are partly based on his perception that 'with all its seeming openness and friendliness to tradition, postmodernism is the most serious threat so far devised' (1992: 299). This view of postmodern consumption settings implies that resorts tend to 'appropriate' misrepresentations of tradition with a view to undermining them. Even if this is true, it should not prevent us from viewing the types of community which arise from such constructs. If, as Lowenthal (1981) has argued, 'nostalgia is an affliction', there may be some grounds for considering resorts as windows into their immediate and regional settings prompting holidaymakers to 'search for meaning', albeit with varying degrees of intensity. MacCannell's (1992) view of resort tourism as a threat may have some validity in the Mamanucas where elements of tradition are very prominent in the 'offer' made to prospective customers.

Wagner (1977) has portrayed package tourism (and by implication island resorts) as disruptive of local social structures. The permissive sexual behaviour of older Swedish female holidaymakers with younger Gambian men is seen as emanating from the 'spontaneity and immediacy of (resort) *communitas*' and by the 'temporary suspension of status differences'. From the Swedish point of view, resort behaviour is a 'ritual of rebellion'. The impact on Gambian social structures is more corrosive in that respect for older members of the community is lost and 'young men and boys were for long periods of time, being socialised by strangers into living in virtually normless *communitas*'. Wagner (1977) did not examine the long-term impact on the local community but brought out some distinguishing features of resort life.

Another useful typology for island resorts is Redfield's (1955) model of 'little communities'. This typology can be used to distinguish between the various large and small Whitsunday and Mamanuca resorts. Redfield's 'face-to-face societies' are exemplified by the smaller, more intimate resorts where all members of the community have a personal relationship with each other. The larger resorts are more symptomatic of mass modern society where individuals relate to only a fraction of all the members of the group.

A number of authors have depicted the modern quest for the 'village in their mind' (Wright 1985; Urry 1990: 97). Urry has commented on the English trend to construct 'new estates in vernacular or rustic style – usually described as villages' (1990: 97). Most of the resorts in the Mamanucas and Club Med in the Whitsundays seek to reproduce a 'village feel', though the Mamanucas with their *bure* configurations are closer to the original Club Med village concept than the Lindeman Island 'village' itself. Wright refers to an 'abstract and artificial aestheticisation of the ordinary and the old' (1985: 230). In such postmodern villages, 'people live in different worlds even though they share the same locality: there is no single community or quarter. What is pleasantly old to one person, is decayed and broken to another' (Wright 1985: 237). His comments, though not directed specifically at resorts, might well be applied to contrasting views of resort 'villages' by visitors and by the local workforce. It is perhaps inevitable that these two groups will have contrasting perceptions. Resorts occupy ground somewhere between the 'ex-primitive' (using MacCannell's term to apply to traditional Fijian society) and the modern. If we apply Tonnies' Gemeinschaft/ Gesellschaft division to distinguish between traditional community and modern society, we can see the island resort as an attempt to recapture some of the attractions of the 'traditional' model, whilst not forgoing the advantages of modernity.

MacCannell sees the absence of an historical dimension as corrosive. 'A postmodern community' is 'a glitzy warehouse for souls. It binds people together by destroying all ties between them. It claims to put them in contact with their innermost feelings, but does not allow them to talk to their dead ancestors' (1992: 94). Brampton is an example of a Whitsunday resort where the destination markets itself as a place where one can 'rediscover oneself' and this is a not uncommon theme of resort advertising generally.

The attempt by resorts to give the appearance of social uniformity is a particular irritant for MacCannell and other sociologists. Critics are apparently not satisfied by the fact that early Club Med philosophy was dominated by an objective of breaking down class boundaries through their use of beads as currency, emphasis on informality and simple accommodation. The 'classlessness' of early Club Med (some have termed it 'bourgeois') had a precursor in the early English holiday camps with the use of 'campers' as a quasi-classless way of describing their clientele. MacCannell's concern is that true class differences are concealed, just as clients from ethnic minorities must conform and be 'transparent'. In practice most of the market segmentation activity undertaken by the Whitsunday resorts is aimed at attracting particular socio-economic brackets within the community. Thus South Molle Island is targeted at the 'lower middle classes' (K. Collins pers.comm.) and Hayman at the top five per cent of earners in the Australian population (King and Hyde 1989b). This practice may certainly have the effect of reproducing some of the social stratification normally found in the suburbs of major Australian cities. The class criticism has been applied to the Queensland resorts most insistently by Craik (1991) who has argued that there is a lack of holiday options for lower income groups. There is no doubt that higher costs are prevalent in the Whitsunday Islands. Nevertheless, it is feasible for low income earners to holiday on the Whitsunday mainland where cheaper accommodation options such as caravan parks are available and ancillary costs are lower. The absence of such options on the islands themselves certainly results in a degree of social segregation which Craik regards as unacceptable. The author's view is that Craik overemphasises socio-economic segregation as a negative, whilst lower-cost holiday options such as caravan parks may in fact suit the holiday preferences of many consumers including certain lower income earners.

TO WHAT EXTENT CAN THE RESORT EXPERIENCE BE DESCRIBED AS AUTHENTIC?

Authenticity may be defined as 'not spurious or counterfeit; genuine, original; certified by valid experience; unquestionably true' (King 1994). Since island resorts are 'consumption places', they do not apparently lend themselves to authentic experiences – in a sense everything that takes place is contrived and has an underlying profit imperative. This is particularly the case where the only permanent island population consists of resort employees. In such situations, the only 'authentic' experience would occur *off* the island, or in instances when residents from neighbouring islands visit the resort. Most instances of this, however, would involve such locals performing for resort guests.

For the purposes of the present study, P.L. Pearce's (1988) more liberal definition of authenticity is relevant. He argues that 'tourists can achieve authentic experiences through relations with people in tourist settings' (1988: 179). He also asserts that authenticity 'can be achieved either through environmental experiences, people-based experiences, or a joint interaction of these elements' (1988: 179). His acknowledgement of both environmental/landscape dimensions and social dimensions gives this view of authenticity particular potency for the current research. Using Goffman's (1959) theory of backstage and frontstage settings, Pearce states that 'a back-front distinction can be made for the authenticity of both the setting itself and the people in it'. Pearce provided nine classifications of tourist experiences. The typology would allow a situation such as the following to be classified as authentic: where a tourist has an opportunity to mix with a member of staff outside working hours and gain insights into his or her life. The fact that the island is not the site of permanent habitation does not necessarily inhibit the occurrence of an authentic experience. Pearce's appraisal enables us to view the physical and social environments from an integrated point of view and to depict guest satisfaction as involving a blend of these two elements. It also implies (though does not state explicitly) that it is useful to think of resorts as communities occupying an overlapping physical and social context.

SOCIAL INTERACTIONS AT RESORTS

As previously stated, a wide range of social encounters and activities take place in resorts. Urry has stated that 'tourists typically move into and through various sorts of public space – such as beaches, shops, restaurants, hotels, pump rooms, promenades, airports, swimming pools and squares. In such places people both gaze at and are gazed at by others' (1990: 140–1) and there is a complex range of expectations about who one would like to look at one and who one would like to look at. He also depicts tourism settings as a 'licence for permissiveness and playful "non-serious" behaviour and the encouragement of a relatively unconstrained *communitas* or "social togetherness"' (1990: 10). Urry has suggested that it is the 'tourist gaze' which differentiates tourism places from everyday places in the way that people look at them (and are themselves in turn looked at). Island resorts are examples of what Pearce calls 'tourist places', that is, places where the predominant function is tourism. This makes the 'tourist gaze' easier to identify. The main criticism of Pearce's approach is that it devalues the importance of linkages between the tourist settings and the local economy. The villages on Mana and Malolo islands, for example, though physically separate from the resorts, inevitably result in human interactions. Whilst the resorts on these islands *behave* as if they are self-contained, offering 'exclusivity' to their guests, they are not physically differentiated. As will be discussed in chapter 8, the resorts exploit the ambiguity between their exclusivity and the freedom of guests to roam at liberty on adjoining – usually public – land. There is a parallel between the exploitation of this dichotomy and the relationship between the water and the land. Thompson (1981) has referred to the coast as a boundary between the land, representing the familiar, the knowable and the controllable and the sea as a terrain which cannot become known or subject to human control. The juxtaposition of swimming pools (a private resource) and the sea (a public resource) seeks to exploit the sense of freedom provided by adjacent public territory, whilst giving the reassurance of private control.

MacCannell (1992) has depicted interactions between local residents and staff as relationships between superiors and inferiors. The only mutual understanding arrived at between locals and holidaymakers may be likened to the torturer saying to his victim 'we now have an understanding' after he has broken the prisoner's spirit. Urry prefers to depict the relationship as one of intrusiveness. Tourism settings are examples of how 'modern society is rapidly institutionalising the rights of outsiders to look into its workings' with visitors wanting to see how local residents live (1990: 8). The author found that the Mamanuca villages and the camps (or 'villages') used to house employees are deliberately made very visible to visitors. Though it could not be ascertained whether staff resented this intrusion into their privacy, it is clear that uncontrolled access by houseguests has the potential to disrupt and offend. When visiting Tavarua Island resort, the author was taken to a 'Fiji village' where the staff live. Whilst the visit was intended to add an extra dimension to the author's trip to the island, the impression conveyed was like a museum visit with human props. As Urry has stated, 'almost everywhere has become a centre of "spectacle and display"' (1990: 93). Given the small dimensions of such islands, the constant proximity of the guests and the absence of privacy from the 'tourist gaze' is a notable characteristic of postmodernism and may be disconcerting to employees. MacCannell (1992) is also critical of the pressure on local populations to 'act primitive for others', being asked to reconstruct a world which never existed.

Island resorts are characterised by explicit public display. Since there are no residents available to photograph, apart from staff, most promotional imagery depicts tanned and scantily-clad holidaymakers wearing bikinis (in the case of women) and swimming costumes (in the case of men). Bodies are predominantly uncovered, exposing various shapes and size of physique normally concealed by clothing. The practice of 'going barefoot' reinforces the image of informality and social fluidity. At Beachcomber Resort in the Mamanucas, the sand from the beach is the material underfoot in the bar area too, giving a physical dimension to informality and to linking the human (buildings) and the natural (the beach).

The Mamanucas adhere to western dress norms in that holidaymakers wear swimming costumes though they are modified by the addition of a *sulu* or wrap-around length of cloth tied at the waist. The addition of the sulu is a concession to the Fijian preference for 'covering up', a legacy of Christian missionaries. Though the dress code contrasts with prevailing norms in the original island communities and the practice of most Fijian resort staff, it is symptomatic of cross-cultural compromise in Fiji. The practice of sulu-wearing has apparently been readily accepted by Caucasians and provides the holiday experience with some local flavour. It exemplifies the introduction of local practices in a sun, sea and sand destination in place of the complete importation of alien cultural practices. Sulus apart, the public display of bodies in both Fiji and in the Whitsundays follows an implicit code and there is some expectation that guests will conform with such norms. Whilst minorities with alternative dress codes (say Moslems or non-urban Aborigines) would not be blatantly excluded from such resorts since the determining principle is the ability to pay, there is pressure to conform with certain behavioural norms, including dress norms.

The selective range of promotional images used to depict island resorts provides a context within which the 'tourist gaze' will take place. Choosing to go about one's business scantily clad involves conforming to a variety of what John Berger has called a 'way of seeing' (1972). Those who do not feel that they conform may feel intimidated by such expectations. Three female Melbourne focus group participants perceived the need to conform to a certain stereotype and to make a public display of their bodies which they felt were not 'up to scratch' and consequently felt deterred from going to island resorts. The view was not contested by other female (or male) participants in the relevant group. The need to be young, tanned, glamorous and scantily clad was seen as a barrier to enjoyment and relaxation for those who did not think of themselves in these terms. The view of resorts as an 'unattainable paradise' may appeal to the image makers, but has unfortunate by-products for certain women travellers. Other groups expressed the more pragmatic view of resort selection, emphasising perceived suitability according to one's activity preferences.

The beach is sometimes thought of as a great leveller (Spearritt 1976: 219). J.D. Pringle in his book *Australian Accent* captured the levelling notion with his comment that 'you cannot tell a man's income in a pair of swimming trunks and the Pacific surf is a mighty leveller' (1958: 198). The steelworkers of Newcastle, NSW have as ready access to the city surf beaches as the moneyed élite. Island resorts, however, are able to engage in a more explicit type of social control. Such segregation is apparent at the time of resort construction where lavish facilities will obviously lead to high prices and to a market enjoying above-average disposable incomes. Influence can also be exerted by controlling the number of packages using discount seats available to particular markets. Hence Hayman Island attracts only limited numbers of families at school holidays (and thus retains its desired market mix) because of the higher prices charged for flights during such periods. The separation of islands from the mainland by water also provides an opportunity for market segregation. At Daydream Island, many daytrippers are less affluent than house guests and many stay at cheaper mainland accommodation. Daytrippers are not admitted to the main resort area and are confined to the beach club. Such a practice on the mainland would involve a more intrusive type of policing than is necessary at Daydream. The influence of packaging of island resorts in major source markets such as Melbourne and Sydney also allows for market differentiation by island. In the Mamanucas the distinctions are less clear cut, since there is only one 'deluxe' property, the recently opened

Sheraton Vomo. Though the others offer differentiated pricing and some appeal to particularly targeted groups (e.g. Tavarua targets surfers), the market is predominantly a middle-class Australian one.

Are the two island groups international communities or a small slice of Australia? The survey and interview results demonstrated some contrasts. The Mamanucas are losing their traditional dependence on Australian market sources as a more diverse international clientele is attracted. The Whitsundays have been *increasing* their dependence on Australians, particularly from interstate. Whilst certain Whitsunday resorts attract substantial international visitation, a number remain steadfastly domestic with the sense of being Australian enclaves.

Whether their guests are domestic or international, one of the challenges for resort management is to create an agreeable 'ambience' in which guests will feel relaxed and at ease (Gee 1988). Most island resorts espouse a philosophy of 'relaxation' for their guests. Recognising this fact, major international hotel groups often separate their hotel and resort divisions with the former emphasising business efficiency rather than relaxation. The achievement of a level of relaxation that suits all client groups is difficult in the islands because of the wide range of preferences, with some clients associating relaxation with unwinding or recovering after a bout of vigorous activity. For others, relaxation means, say, sunbathing or reading. Somehow a resort must cater for as much of this wide spectrum of preferences as is practicable and in advertising terms, believable. Part of the process of ensuring that guests have a rewarding experience socially is undertaken in the process of marketing. The language of advertising is used to convey messages that receivers and potential consumers use to make judgements about what can be expected at the resort. Since island resorts are artificial creations, a social atmosphere or ambience must be engineered. Such engineering takes different forms in the various resort islands and elicits very different responses.

In the Whitsundays, management at Hook Island Wilderness Lodge and Palm Bay Hideaway emphasised the 'traditional' Whitsunday approach. According to Tall (pers.comm.) Hook is 'getting back to what the resorts used to be like twenty-five years ago'. These properties have more in common with the smaller Mamanuca resorts than with their bigger counterparts in the Whitsundays. Nostalgia for a previous era is not necessarily effective from a marketing point of view and Palm Bay is apparently more successful in attracting international ecotourists than mainstream domestic consumers. In advertising 'home-style' cooking, Palm Bay may appear as old-fashioned rather than as reassuringly nostalgic amongst the increasingly gourmet-minded Australian consumers.

The resort managers at both properties claimed that ambience did not involve manipulation by management. At Hook 'It just happens. This is not a Butlins Holiday Camp. We have the odd night observatory run or game of volleyball and people get together for a beer after a day's snorkelling, but we don't want too much activity. We don't run around with a megaphone' (C. Tall pers.comm.). The view that 'it just happens' is debatable, given that Hook Island Wilderness Lodge is as artificial a creation as any other Whitsunday resort. It does suggest a lower level of institutionalised intervention though, than is present at the larger, more formal resorts. Like Palm Bay, Hook's anti-institutional approach has more in common with the Mamanucas than with the larger, more typical Whitsunday resorts.

A more institutionalised approach was evident at the other Long Island property, Radisson Long Island Resort. Social engineering was deliberately practised to bring about increased guest spending. According to N. Kelly (pers.comm.), the food and beverage activities at Radisson Long Island were aimed at creating ambience through special themes. 'Tonight it's a Mongolian Barbecue where guests can eat as much as they like. This encourages them to stay around the bar afterwards. If people have a restaurant dinner they tend to go to their room afterwards.' This is an obvious example of resorts as 'consumption places' with events specially tailored to prompt consumer spending. The author noted that the walls of the general manager's office were lined with the several volumes that make up the *Radisson Operations Manual*, a company-wide publication which senior management see as a reference point for all properties.

Social engineering is most consciously undertaken at Club Med under the label 'the Club Med Spirit'. According to the Chef de Village (B. Giampaolo pers.comm.), 'we try to introduce couples to couples'. Interaction is also encouraged between staff and guests: 'We try to enable guests to have fun and to share a lot of things with the *Gentils Organisateurs* (GOs) and *Gentils Employés* (GEs)'. The fact that GOs are deliberately selected for their outgoing personalities ('they are always smiling, happy and close to the guests'), sociability and for their proficiency in performing in sport and at floor shows is an indication that 'atmosphere' will consciously be generated. Worldwide, Club Med emphasises the quality and diversity of its activities and of its food, both of which are seen as 'shared experiences'. On the day that the author visited Club Med, a large picnic had been organised to the other end of the island. Club Med are 'not so good at room service', according to the Chef de Village. As with the style of food at Palm Bay, this philosophy may be incompatible with the expectations of Australian consumers.

In the Mamanucas, all managers identified the friendliness of Fijians as a key element in marketing their business. Resort managers (predominantly Caucasians) described Fijians as better than Indians in 'front-line' positions, that is, those roles involving direct interaction with consumers. One manager (H. Steinocker pers.comm.) stated that 'for engineering and in the kitchen, the Indians are better, but as front-liners the Fijians are better'. Though they do form an essential component of the workforce, there are relatively few Indians in interactive roles. The frequency with which type-cast views of Indians and Fijians are expressed must be of concern to applicants from either social group for positions which do not conform to that stereotype. Whilst racial stereotyping is present in many aspects of Fiji society, it is particularly blatant in an image-based industry such as tourism. Whilst the government-owned Air Pacific uses both smiling Fijian and smiling Indian faces in its promotions (see figure 7.1), promotions undertaken by tour operators and resorts almost always depict smiling Fijians occupying interactive roles. When assessing travel brochures, one would scarcely be aware that Indians make up almost 50 per cent of Fiji's population. Smiling Fijian faces are commonly used as the centrepiece for brochure covers. Australians may return home from the Mamanuca resorts unaware that Fiji is a multiracial and multicultural country. This attitude may suit the implicit political agenda of the current Fijian-controlled government but may result in the dissemination of distorted, unrepresentative images of Fiji overseas. One of the dilemmas of adopting a narrowly island base to the Mamanucas tourism region is that only Fijians are long-time residents in the islands. The nearby mainland towns of Lautoka and Nadi, on the other hand, have substantial Indian populations.

The philosophy of service as practised in the Mamanucas has parallels with the Whitsundays but there are some marked differences. Sometimes the public relations value of the 'friendly Fijian smile' is seen as meriting some extra tolerance towards inefficient methods. The Castaway manager (G. Shaw pers.comm.) remarked that 'the food and beverage, front of house and housekeeping areas are crucial. It doesn't matter if the housegirls do relatively few rooms – it's most important that they talk to the guests'. This is not social engineering but it is certainly an example of management adapting expectations of service to the local culture.

Shaw's comments are the closest any manager came to espousing a distinct philosophy of service. For most managers service involves bringing in expertise from outside (for example diplomates for the Fiji Hotel and Catering School) and blending this with the local landowning staff recruited as an obligation to the locality. With local staff, inculcating technical skills is seen as particularly important amongst recruits with little previous experience outside the village setting. Shaw (pers.comm.) talked about the need to 'achieve a balance between the genuine friendliness of the Fijians and efficiency', though he acknowledged that the level of efficiency 'has to rise'. He described service in his resort as 'lacking the efficiency and sharpness of service in Honolulu'. Despite the qualifications expressed, however, managers in the Mamanucas viewed the Fijian attitude as a positive.

At a number of the resorts staff social interaction with guests is encouraged, notably at Plantation, at Beachcomber and Tavarua. According to Kalo Tuai (pers.comm.), 'the staff play rugby and volleyball with the guests

Figure 7.1 Air Pacific staff
Multicultural Fiji: the Air Pacific version is not reflected in the promotion of Mamanuca resorts
which invariably depict smiling Fijian faces.
Source: Ernest Dutta, Director of Marketing, Air Pacific (courtesy of Air Pacific)

and they mix freely in the evenings, for example, dancing and drinking kava'. In most of the resorts, staff are not allowed to drink alcohol. The availability of the non-alcoholic (though mildly stupefying) kava as an acceptable surrogate for alcohol in the Fiji resorts is probably a positive in that it makes prohibition appear a less draconian policy. The philosophy of interaction is also strong at Tavarua. According to Jon Roseman (pers. comm.), 'we really like to integrate our staff with our guests and with twenty-four guests and twenty-seven staff that means one-to-one correspondence'. Social interaction is seen as an opportunity to communicate elements of the local culture to the guests. A weekly Fijian night is staged where guests are introduced to Fijian traditions and to kava and Fijian food is cooked on a couple of evenings each week. According to Roseman (pers.comm.), 'the staff manager's great-great-grandfather was buried here in a beautiful grave surrounded by rocks. We like to take our guests over there and explain the background story to it. We like to capture the whole feeling of the place for our guests. We don't want them to just have the surfing experience and then leave'. Relating cultural heritage to the individual experience of staff is seen as integral to the resort operation.

Though the scale of operation is larger and consequently the approach is less personal, a similar philosophy was espoused by managers at Beachcomber Resort. Staff are encouraged to communicate their own cultural traditions to guests, including their language, the kava ceremony, basket- and hat-weaving and sulu-tying. Kalo Tuai (pers.comm.) claimed that staff 'start to remember things that had been lost'. Though not stated explicitly, there was an implicit indication that staff who were not interested in sharing their cultural traditions would be regarded as less suitable. Staff are certainly under pressure to 'perform', though this is not necessarily a negative.

There is a consumer and travel industry perception that the Fijians are 'losing their culture' (Stollznow Research Pty Ltd 1992: 148). This negative was less evident in the Mamanucas where the blend of local staff and staff drawn from other parts of Fiji appears to work well. Fijians have a strong tradition of communal obligation and the sharing of cultural traditions with other staff and with the guests themselves appears to be readily accommodated. The setting may also be an advantage in that 'the Mamanucas are better off (re: staff) than the mainland properties. There's an island atmosphere. Fijians are normally happy to be out on an island – it's partly a freedom thing' (A. Reed pers.comm.). Jim Saukuru (pers.comm.) also stated that Fijians enjoy working in the Mamanucas. The smaller geographical dimension of Fiji means that workers from other parts of the country maintain close family connections with their region of origin and may have family connections with other staff. The author led a group of University of the South Pacific tourism students on a field trip to the Mamanucas and was struck by the large number of family connections between Fijian students and resort employees. In Australia the prospect of (usually short-term) work in the Queensland islands is an appealing prospect for young Australians, though the weaker communal ties and vast scale of Australia mean that links with home are more tenuous. Although young Australians may deliberately seek employment on the islands to escape home ties and obligations, the real isolation from family and other support networks suggests that engendering a sense of community and mutual support amongst staff may be important to achieve a contented and satisfied workforce.

THE STYLE AND QUALITY OF SERVICE

Consumers perceive Fiji as a 'friendly' destination for the interactions it offers amongst guests and between guests and resort staff and/or local residents. The *bula* (welcome) greeting and friendly smiles of the Fijian staff are emblems of such perceived friendliness. The Whitsundays are perceived as offering a more perfunctory style of service – certainly more professional and more efficient, but less friendly. One telling comment expressed in a focus group was that 'Australians are more friendly to one another when they are travelling overseas' and it was stated that in the Whitsundays guests tend to 'keep themselves to themselves'. Focus group respondents perceived that water-sports enthusiasts and 18–35's would be better placed to enjoy the social dimensions of the Whitsundays than other targeted groups. The Australian resorts such as Dunk Island which attract a large international clientele were seen as offering the best social experience.

With the exception of villagers on Mana and Malolo islands and the guests themselves, everyone in the Mamanuca resort islands is involved in some form of tourism service. All those who are not paying guests of the resorts are, in a sense, playing a role (or 'performing') and any interaction between a guest and a 'local' is derived from a commercial rationale. Interactions cannot be divorced from the style of service offered by the resort. On islands where accommodation is provided for the spouses and children of employees, some less commercially-based interactions may take place.

A variety of service levels are on offer within the Whitsundays, though the room capacity of the 'resorts' at the lower and less formal end of the market is very small. At Hook, the service style is described as 'very basic' with more emphasis being placed on friendly as opposed to 'professional' service (C. Tall pers.comm.). The typical meal at Hook is 'home cooked, for example, roast lamb, corned beef and mashed potatoes'. For those seeking extra atmosphere 'we set aside six tables inside with tablecloths on them, knives and forks and put a candle on the tables'. The implication is that through the placement of some largely symbolic representations such as tablecloths and candles, a certain sophistication of ambience can be achieved. Tall's perception appears to be that the juxtaposition of the highly informal with the quasi-sophisticated adds colour and character to the resort as well as enhancing the visitor experience. It is reminiscent of an older-style Australia, less influenced by current international trends.

The most distinctive service style was evident at Club Med. According to Bernard Giampaolo (pers.comm.), we 'care more about eating and about daytime activities than about rooms and restaurant service. Our rooms are very nice but they are not very luxurious'. Quality service is also seen as a part of the provision of activities: 'service means that if you want tennis then it is made available immediately, if you want to play golf, then you can always do so, since the equipment and the instructor are always there'. The 'international ambience' is arrived at by the 50/50 mix of Australian and international GO's (all GE's are Australian). 'The head of beach activities is Tahitian, the chief cook is French, the chef de village (me) is Italian, the heads of sport are Moroccan and Canadian.' The author's observation was that the international ambience was quite genuine at Club Med, albeit with an Australian flavour. It had the strongest international feel, with the possible exception of Hayman. Despite the remarks of the Chef de Village, the guest rooms at Club Med did appear relatively luxurious and well appointed, certainly far removed from the simple hut-style accommodation originally promoted by Club Med at its French 'villages' in the 1950s.

Hayman has deliberately transformed its style of service in order to reposition the resort. As was the case with Hamilton Island, Hayman's initial marketing focused on exclusivity, with a deliberate projection of élitism. Speaking shortly after the redevelopment of Hayman, then Managing Director of owners Ansett Transport Industries, Sir Peter Abeles stated that 'the new Hayman is expensive and beyond the reach of the average holidaymaker' (King and Hyde 1989b: 205). Unfortunately for Hayman's owners, the target market was smaller than anticipated. The current General Manager of Hayman (T. Klein pers.comm.) describes the earlier approach as 'stuffy, formal, very exclusive and very European'. According to Klein the style has now become 'more Australian, more relaxed, more fun and more active. There is now more emphasis on the domestic (market) and on the family'. Socialisation at the resort had been a problem – 'a problem for Hayman had been that people were not sure how to behave. Can I walk around in shorts and thongs? What do I wear? Do I need to wear a jacket? We have tried to change all that'. In the 1980s, Hayman failed to conform to consumer expectations about what constituted island relaxation.

A revealing aspect of Hayman's repositioning has been its deliberate encouragement of a distinctively Australian atmosphere. The resort's Polynesian Restaurant was changed to an Australian theme in 1993. According to Klein, the impetus came from international rather than from domestic demand and involved the incorporation of, for example, crocodile on the menu. I was concerned about the response of the local market to the inclusion of kangaroo and emu, but the change has been extremely well received. We have received one or two negative comments but the predominant domestic and international response has been very positive.

(pers.comm.)

The restaurant change is symptomatic of the increasing internationalisation of Australian tourism resulting in opportunities for Australians to view their own country differently. Another dimension of creating a distinctly Australian environment has been that 'there is a lot of wildlife that has been brought into the national park from other parts of Australia'.

Remoteness can be a barrier to creating an ambience based on efficient but sensitive service. General managers acknowledged that the relative isolation of the Whitsundays has impeded the recruitment and retention of quality staff. Most interviewees claimed that the increasing attention being devoted to training and service issues is a fairly recent phenomenon but that the effort is yielding positive results. According to Hutchen (pers.comm.), the quality of service 'is a problem'.

At one end of the scale, Hayman aims to be 'world best'. According to Tom Klein (pers.comm.), 'We [Australians] have come a long way over the past fifteen years. It [service] is not a problem for the international market. We can compete with the best in the world'. Cost is perceived as the other main problem relative to developing country tourism destinations: 'Labour costs in Australia are high. 460 staff live here year round. We have no part-time or casual staff so fixed costs are high. The typical duration of service is eighteen months and two to five years at management level'.

Club Med has the most international staff of the Whitsunday resorts with its global policy of recruiting over half of its GO's from overseas. Giampaolo (pers.comm.) claimed that 'In Europe expectations of customers are ten times higher' than in Australia, though Mahony has argued that 'Five years ago if a domestic customer had a lousy meal or service, they would probably have put it down to experience. Now they would knock down the door and let you know about it.' The latter comment points to an increasing alignment of Australian expectations with standards prevailing elsewhere. The overall impression is that resorts which attract a high proportion of international guests are being pressured to provide a higher-quality service to meet their higher requirements. Higher expectations amongst Australian clients are also placing upward pressure on standards, though the response of the different resorts to such altered behaviour is patchy.

Kelly cited the main differences between the Radisson Long Island property with the Radisson in Vanuatu as the different levels of community integration and cost. 'In the Whitsundays you have less community infrastructure. This means that staff get quickly bored and want to move on. They are very itinerant with no commitment to the place.' He also pointed to the opportunity for higher staffing levels in the South Pacific: 'In Vanuatu we employed 180 staff for an equivalent property where we employ seventy-five here.'

In the Mamanucas, interaction between the resorts and the nearby communities impacts upon service quality. One might anticipate a higher level of interaction where there is a Fijian village on the island. In fact the bond between resorts located on previously uninhabited islands and the relevant landowning community are often very close and amicable. This point was clearly articulated by Jim Saukuru (pers.comm.) who stated that 'the fact that the resorts are so interwoven with the villages and that they are so close to us is a real advantage. Guests really appreciate being able to visit the Fijian villages in the Mamanucas – it gives them a real taste of the culture and lifestyle.'

As previously mentioned, Malolo island has been continuously occupied by a Fijian settlement. Most others have semi-permanent employee accommodation in a 'village' setting. At Sheraton Vomo, the relevant village was described as 'our other hotel' and at Tavarua as 'Fiji village'. In most cases, resorts have a binding agreement to recruit at least half of their labour force from the relevant *mataqali* or landowning unit. Some *mataqali* staff come from villages on adjoining islands (e.g. in the case of Matamanoa and Tokoriki) and others from the mainland (Beachcomber and Treasure island resorts typically source staff from Viseisei village).

Resort managers described the overall relationship with the *mataqali* as 'cordial', though Helmut Steinocker (pers.comm.) complained that village recruits who might have no high school education nor any training were able to claim full union rates after three months on the job. Since villagers would be typically unaware of the 'western' (i.e. holidaymaker) way of doing things, he claimed that the lead time was unrealistic. Such comments, though opposed by some other managers, underscore the dilemma of using western value systems in the employment of villagers used to a semi-subsistence existence.

The remoteness of the 'western side' of Fiji (i.e. including Nadi and the Mamanucas) from the main urban area of Suva was cited as a problem for recruitment. A major Nadi-based operator explained the appointment of predominantly Indians to clerical roles was because of the dominance of the 'village-based' lifestyle for the Fijians in Nadi, unlike the more sophisticated urban lifestyle of Suva (R. Whitton pers.comm.). This predominance of Indians in Nadi is reversed in the Mamanucas where the bulk of employees are Fijian. The remoteness of the resort islands from the relevant hospitality training facilities is clearly a challenge in both the Mamanucas and the Whitsundays.

The presentation of Fijian culture and the 'friendly Fijians' is treated as a core attraction by most property managers. The manager of Mana (K. Palise pers.comm.) stated that 'We have a lot of cultural activities . . . including full floor shows three nights a week. We have mekes and fire-dancing, kava ceremonies and woodcarving, not karaoke or discos'. He regarded the beaches, water and activities as equivalent to those offered by the Australian resorts and culture as the main point of differentiation. Like Saukuru, Palise placed a heavy emphasis on the idea of the resort as an integrated community and cited the fact that villagers on the island actively welcome visitors. In view of the major resort expansion envisaged for Mana Island, it will be a challenge to maintain the integrity of this concept.

Local community relations are seen as maintaining cordial relations with the relevant chiefs and being perceived as good employers. In the case of Beachcomber, 'we keep close contact with the chief . . . [the *Tui Vuda*] . . . he comes into this [the Lautoka] office each week. We look after him with little personal things such as phone calls. He sometimes seeks our assistance with a village happening' (A. Dreunimisimisi pers.com.). These comments point to the informality of the relationship. This informality allows for flexibility but leaves open the potential accusation that chiefs may gain personal benefits at the expense of other landowners. The Assistant General Manager of the resort, Kalo Tuai is the son of the chief and reached his present position after occupying a variety of subordinate roles at Beachcomber. Not surprisingly in view of his local credentials, Tuai sees himself as a key agent for bringing about a higher profile for cultural activities within the resort. Maintaining the appropriate balance between building relations with 'commoners' and with the Fijian chiefly hierarchy is a source of constant tension in Fiji's national politics and is an issue at regional level also, though it was not a subject which interviewees were interested in discussing.

As previously indicated, most front-line staff are Fijians. Fijians respond to western-style incentives in their own way. In a formalised work environment, family obligations such as attendance at funerals may account for the absence of a Fijian employee over several days (T. Trustler pers.comm.). Conspicuous accumulation of wealth is not regarded favourably and possessions are shared with others. Monetary rewards, though welcomed, are less potent incentives for individuals where there is an expectation that much of the surplus will be shared with the group. Such cultural differences are a source of intrigue for visitors who choose to take an interest in such things, but are a challenge for managers.

The status quo between villages and resorts is periodically upset. For the most part, the resorts seek to maintain control over visits to the villages, ostensibly on the basis that too many staring eyes can disrupt the life of the village, particularly where guests are ignorant or misinformed about protocol. The same concern is expressed by the Fiji Department of Tourism which promotes 'controlled tourism' including 'safeguarding traditional values and norms' (Fiji Department of Tourism 1992: 18). The negative from a villager's point of view is that the high rates charged per bure (e.g. F$350 per night for a deluxe bure on Mana) do not penetrate to village level, except as watered-down wages and lease payments. It is tempting to offer the local equivalent of a bure at a much lower rate, especially when the style of accommodation for locals and guests is superficially similar. NLTB officials acknowledged that tours have been organised from the mainland promising the opportunity to 'stay in a village' whilst enjoying all the amenities of a resort (T. Waqaisavou pers.comm.). The village on Mana Island was cited as an example of where such practices have occurred. The dilemma for resorts is that where the Fiji authorities are starting to emphasise the 'authentic Fijian experience', the response by holidaymakers and by villagers may be to bypass the resorts altogether.

Naturally the resort managers do not welcome this form of competition. The resort facilities and infrastructure were privately installed and the accommodation rates charged attempt to achieve a return on that investment. In these circumstances the 'informality' which determines the boundaries of an integrated resort are tested. So far such incidents appear to be relatively isolated, but the perception that tourism is a 'white man's industry' can spark such consequences. Another negative mentioned at the Fiji Department of Tourism is a periodic law and order problem with 'landowners harassing expatriate managers to extort money' (M. Gucake pers.comm.). Overall the respondents' impression is that such confrontations have been on a very small scale. Nevertheless, caution is needed in ensuring that image building concerning the friendly Fijian and of tourism living in harmony with community and environment is not overdone.

Another social dimension of life in the Mamanuca islands is the near elimination of the subsistence economy and its replacement by a tourism cash-based economy. The Director of the National Trust for Fiji (B. Singh, pers.comm.) related that trust activities are severely constrained in areas such as the Mamanucas because of an inability to recruit volunteer labour to provide visitor and custodial services at heritage sites. In other areas rangers can be appointed on an 'honorary' basis with payment in kind through clothing and fuel. Tourism projects in areas of Fiji where alternative employment opportunities are available face similar problems. This is a dilemma in a developing country such as Fiji where public funds are limited, particularly in fields such as conservation which are typically regarded as luxuries by government. (The 1994 budget for the National Trust of Fiji was a paltry F$80,000.) In 1993 the Minister of Housing and Urban Development is reported (B. Singh pers.comm.) to have promised F$940,000 towards Fiji's first national park at the Sigatoka sanddunes on Viti Levu. By the time the proposal went to Parliament, the proposed sum had been reduced to F$150,000.

THE RISE AND RISE OF THE FAMILY

For resorts attempting to attract families, the provision of activities for children is essential. The emergence of childcare centres as a key marketing platform for the Whitsundays resorts in particular typifies a return to the family market – but with a difference. Whilst not wishing to forgo other social groups (such as honeymooners), resorts are attempting to bring together a diversity of groups in a confined space. This strategy must face the prospect that certain guests may be deterred. The solution chosen by the resorts has been to develop facilities to segregate the 'rogue element', i.e. children (See figures 7.2 and 7.3). Childcare has the combined appeal of providing a practical solution to the dilemmas of market segmentation in a confined space, of being an inducement to parents and of prompting children to urge their parents to return to the same island for future holidays. Increasing childcare provision is a reflection of broad societal trends such as the increased percentage of working mothers and the increasing demand for childcare as a right and not just a privilege. Childcare provision is not a new concept for resorts since the early English holiday camps extended their appeal to families by offering diversions and care for children, thus allowing parents an opportunity to enjoy their own holidays unencumbered. The present-day 'rediscovery of the family' by resorts signals a decline in the practice of gearing a whole island to a single target market (Great Keppell Island sought the singles market and Brampton the honeymooner market). The pressure is on for resorts to provide high-quality, child-oriented facilities and services and other inducements aimed at the family market.

Daydream, Hayman and Club Med have set a very high standard for children's facilities. The larger scale of the Whitsunday resorts certainly allows for a greater degree of organisation than in the smaller Mamanuca resorts. In comparison, the Mamanucas may be relying excessively on the well-established reputation of Fijians for dealing effectively with children. Some investment may be needed to ensure that the quality of children's facilities matches what is on offer in the Whitsundays. The Whitsundays are well placed to reinforce their image as a family-oriented destination.

To what extent do managers in the two resort groups perceive that they can learn from the experience of their counterparts in the other groups? The Mamanuca managers expressed more interest in learning lessons from the Whitsundays than vice versa, acknowledging the fact that the Whitsundays are currently perceived as effective marketeers whilst the performance of the Mamanucas has been deteriorating. Their own bigger scale of operation means that Whitsunday managers do not perceive the Mamanucas as offering much of a business threat.

We return to the issue of whether the two island groups can be regarded as 'communities'. The Whitsundays can only be classed as communities to

THURSDAY 4 NOVEMBER 1993

WELCOME TO CLUB MED LINDEMAN ISLAND

PICNIC DAY TODAY

Evening Dress Code: Jeans, shirt & tie

7.30	Restaurant Welcome
7.30–9.30	Breakfast (Main Restaurant)
8.00	Golf Lesson (Beginners)
8.00	Tennis Lesson (Intermediate)
9.00	Stretching Class (Theatre)
9.00	Golf Lesson (Intermediate)
9.30–11.00	Late Breakfast (Nicolson's Restaurant)
10.00–12.00	Archery (Lessons on request)
10.00	Departure for the Picnic from the Jetty
10.00	Gym/Toning – Low Impact (Sports Centre)
10.00	Golf Lesson (Beginners)
11.30	Water Exercises (Main Pool)
12.00	Orientation Meeting (Theatre)
12.00	Pool Games (Main Pool)
12.30	Sports Information (Entrance to the Restaurant)
12.30–13.45	Lunch (Main Restaurant)
13.00–15.00	Late BBQ Lunch at "The Top" (Sports Centre)
13.15	Entertainment Games (Main Bar)
14.00–14.10	Village Dance (By the Main Pool)
14.30	Golf Course Clinic (Meet at the Sports Centre)
15.00–17.00	Sailing Trip on our catamaran "Seawind" (Sign up at the Sailing Shack)
15.30–17.00	Archery (Lessons on Request)
16.00	Volleyball (Indoor Court)
16.00	Golf Chipping Tournament (Meet at the Sports Centre)
17.00	Aerobics – High Impact (Sports Centre)
18.00	Stretching (Indoor Court)
18.30	Orientation Meeting (Theatre)
18.30	Music around the Bar
19.00	Radio Games (By the Bar)
19.30–20.45	Dinner (Main Restaurant) – Country/Western And Pastry Buffet
19.30–20.30	Dinner (Nicolson's Restaurant) *Please note that reservations can be made for tonight's dinner at the Guest Relations Desk located by the Bar
21.15	Dance Floor (Theatre)
21.30	Showtime "Plastic" (Theatre)
22.30	Crazy Signs & Dance Floor (Theatre)
22.45	Putting Tournament (Around the Pool)
23.00	Sillhouettes (nightclub) opens.

Note: Please see the boards at the 'Sailing Shack' for information regarding times for watersport activities and lessons. *HAVE A NICE DAY* . . .

Figure 7.2 Club Med picnic day
Wall to wall activities at Club Med.
Source: Nigel Stoker, Managing Director, Club Med (courtesy of Club Med)

MINI-KIDS CLUB

MINI CLUB (4/7 YEARS) DAILY PROGRAM

Welcome	9.00hrs
Pool Games	10.00hrs
Outdoor Activities/Arts and Crafts	11.00hrs
Lunch in the Main Restaurant	11.30hrs/(*12.15hrs)
Quiet Times/Stories	12.15hrs/(*13.15hrs)
Depart for the Beach	14.00hrs–14.30hrs
Return from the Beach	15.30hrs–16.00hrs
Snack Time	16.00hrs
Pool/Special Event	16.30hrs
Shower Break	17.15hrs
Dinner in the Main Restaurant/Evening Activity	18.15hrs
Meet Parents in the Theatre	21.00hrs
(*According to the number of children)	

KIDS CLUB (8/12 YEARS) DAILY PROGRAM

Welcome	9.00hrs
Archery Lesson	9.30hrs
Sailing/Pool/Snorkelling Lesson	10.30hrs
Lunch in the Main Restaurant	11.15hrs/(*12.15hrs)
Outdoor Activities/Rehearsals	12.00hrs/(*13.15hrs)
Windsurfing Lesson/Snorkelling or Beach Games	13.30hrs
Snack Times	16.00hrs
Golf Lesson/Tennis Lesson	16.30hrs
Shower Break	17.15hrs
Dinner in the Main Restaurant/Evening Activity	18.15hrs
Meet Parents in the Theatre	21.00hrs
(*According to the number of children)	

Boat rides, bush hikes and other Special Events as advertised on the Information Board at the Mini/Kids Club.

This schedule is flexible and subject to change according to tides and weather. Each child needs to bring sunscreen, swimming costume, hat and shoes.

Figure 7.3 Club Med Mini-Club
And wall to wall activities for the kids too. Most parents are happy to have their children entertained, but their own need for entertainment is more varied.
Source: Nigel Stoker, Managing Director, Club Med (courtesy of Club Med)

the extent that they conform to Foster's (1986) 'short-lived society' model for cruise ships. The context of the resorts that is evidenced through the range of organised activities offered and by promotional material, is dominated by the natural environment and the setting. The historical connections between people and that environment are scarcely mentioned. The workforce and particularly those in guest interface roles are largely itinerant and from outside the Whitsundays region. Social context is driven by the resort and not by a sense of Whitsunday community. The Mamanucas are also a 'short-lived society' but the sense of community is stronger, exhibiting a reasonable level of integration with Fijian villages, including those designated for resort staff. Even in the latter case, the co-existence of two distinct settlements on the same island – a staff and a guest enclave respectively – provides a greater sense of context. Another dimension of the integration of tourism into a community setting is the sharing of Fijian cultural practices. Participation in the kava-drinking ritual requires no skill and is apparently welcomed by the Fijians. The relative authenticity of the experience allows a blend of performance and participation which appeals to the postmodern palate. The islands accommodate a curious type of community consisting of local Fijian residents, the incoming workforce and the guests. Lower staff turnover is experienced by the Mamanucas than the Whitsundays. The result is still a 'tourism culture', far removed from traditional Fijian concepts, but it incorporates an additional social dimension for guests and importantly enables local Fijians to maintain the communal linkages essential to their identity. The relatively small scale of most developments has contributed to a communality of spirit. In the Whitsundays, the larger scale of the resorts and more formalised layout of the accommodation makes such a sense less feasible.

Very obvious differences are evident in the style and conduct of staff/guest interactions in the two island groups. In the Whitsundays an increasing 'professionalism' of service style is evident, according to resort managers and travel agents, though it is still perceived as less than excellent by consumers who also perceive it as relatively unfriendly. Certain resorts (most notably Hayman) are adopting a more deliberately 'Australian style' for their service. Ironically the pressure for this is coming not from Australian consumers but from international visitors who want to experience a taste of Australia. Australian consumer responses indicate a preference for resorts which attract a mix of nationalities. This preference has important implications for the creation of a social setting or sense of community. With greater interest by overseas visitors in Aboriginal issues, there will be pressure to integrate Aboriginals and elements of their culture in the Whitsundays area. The emergence of a stronger regional marketing focus by the WVB also points to increased linkages between the islands and the mainland area of the Whitsundays. This may have important ramifications for drawing local residents into the operation of the resorts.

The social practices of visitors in the two island groups are similar in that all islands cater to a mix of guests each seeking a different mix of activity and relaxation. Even Tavarua, which is dedicated to an energetic and demanding sport, welcomes non-surfing partners as guests. Relaxation is a key element in all of the resorts, which creates a distinct 'atmosphere' recognised by the author in all resort islands in the two groups. Another aspect of relaxation is informality. As in Foster's cruise, there is a less pronounced emphasis on social status than in metropolitan areas. Discussion amongst guests appears to avoid the topic of occupation (with its status overtones) and place greater emphasis on previous travel experiences. Where guests do wish to ascertain occupations and social status, they will tend to do this through a process of networking rather than by asking direct questions which could be seen as a breach of the code of informality. As on cruise ships, resort guests quickly respond to a particular mode of resort behaviour that makes life bearable and possible.

Part of a resort island's appeal lies in the intimacy that it appears to offer. This fact is recognised by a number of the larger resorts which proclaim the opportunity to get away on one's own (usually with a partner) to a secluded part of the island. It is clear that the Whitsunday resorts have to work harder to sustain a sense of intimacy, because of their larger scale of operation and more institutionalised style (in the sense that most are managed by international chains). The consumer perception that the Whitsundays are not 'friendly' is an indication that the creation of effective social environments has not been successful. In contrast, the Mamanucas have managed to maintain their image of friendliness and social cohesion. This is partly a matter of scale of development with bure-style accommodation providing a stronger sense of 'community' (albeit somewhat contrived) than the standard motel-style wings more typical of the Whitsundays.

Socially, resorts may need to be more adventurous if they are to maintain their appeal relative to the increasing number of (often simulated) competitors. Island resorts can draw upon the tropical island myth as an additional lure, but must be cognisant that technological development can enable mainland resorts to offer a wider range of facilities and landscapes. Such landscapes are often better served by a *range* of expenditure and recreational options than the islands will ever have.

8

RESORT LANDSCAPES

In the literature of analysing tourism places, the term 'environment' and associated words such as 'ecotourism' are popular whilst landscape is not. Why does this chapter refer to landscape rather than the more fashionable environment? Landscape is a term which acknowledges the key roles of sociocultural dimensions and imagery in tourism. This is particularly the case when we take an historical view of the 'discovery' of the Pacific and of the ways in which interpretations of landscape have been propagated. It is particularly important that these sociocultural dimensions be acknowledged. One of the contentions of the present thesis is that concepts of how environment relates to tourism have been too divorced from the human dimension and have attempted to hide behind a sometimes quasi-scientific veil. This chapter takes a particular interest in the way resorts 'package' the landscape and bring about a particular and institutionalised 'way of seeing'. Such concepts draw together imagery, social dimensions and landscape. The act of consumption *en masse* becomes 'place creating' and makes the landscape and social dimensions inseparable.]

This chapter explores landscape from an interdisciplinary perspective, drawing upon art history, sociology, anthropology, architecture and aesthetics, as well as from geography and environmental science. An attempt is made to show how the landscape of tourism is currently depicted and what opportunities exist for resorts to depict landscape in an appropriate and sustainable fashion that also has market appeal. The chapter examines the physical dimensions of the resorts themselves and of their surroundings.

A variety of theories are related to resort landscapes including Sack's (1992) interpretation of 'consumption places', Urry's (1990) view of resorts as a manifestation of postmodernism and Ayala's (1991b) theory of 'resort landscape systems'. The physical dimension of resorts is evaluated from the point of view of public versus private space, architecture, aesthetics and style, appropriation and exclusivity, and functional specialisation. The attitudes of travel agents, resort managers and consumers towards a range of landscape issues are analysed. The approach of resorts to their landscapes is a significant issue for resort managers because of changing consumer attitudes and because of the rise of competition from non-traditional sources. An increasing interest in knowing more about the environment is evident amongst holidaymakers and new technology has allowed tropical resort landscapes to be recreated in unlikely locations.

What do we understand by the expression 'creating landscapes'? It is obvious that the twenty-one resorts of the Whitsundays and Mamanucas are 'creations', many of them the result of entrepreneurial activity by an energetic individual or group. The resort buildings are one physical manifestation of an idea, concept or creation. The creation or concept inevitably involves an element of linkage (not necessarily positive or symbiotic) with the immediate physical environs. As we have seen in the earlier chapter on imagery, ideas about and associations with the various resorts are constantly transmitted through promotional activities. Such imagery is created, recycled and then recreated. Finally resorts have an ongoing relationship with the physical environment which amounts to a need to define and sometimes redefine context. Pollution, designation of protected areas, penetration by tour groups and the application of themes to particular landscapes are all forces in the evolution of resort landscapes.

The present chapter title deliberately avoids the label environment as being too broad. To a sentient being, for example, the word environment may describe *anything* external to that person, including other persons. The term of landscape brings with it a rich literature from the visual arts considered particularly important for consideration of the imagery of the South Pacific islands. One cannot divorce island concepts from resort concepts. The landscape concept also allows for a better description of the relationship between the resort site, the resort facilities and their surroundings, though the present chapter also covers practical issues of managing the built and natural environment such as waste management, energy management and site interpretation as well as the study of landscape systems.

Emphasising the cultural dimension, Daniels and Cosgrove (1988: 1) have argued that 'a landscape is a cultural image, a pictorial way of representing, structuring or symbolising surroundings'. They take the view that artistic form and style are insufficient as a complete explanation. Davidson (1994: 6) emphasises the role of human interaction in landscape. To him 'the term landscape as opposed to "land", implies an interrelationship with people – a way of seeing or shaping the land, a framing of it in some social or aesthetic context'. This approach contrasts with the view that landscape can be entirely natural, untouched, objective and defined in purely scientific terms. Geographers have often placed heavier emphasis on the more 'objective', 'scientific' dimensions of landscape. Calder (1981) for example describes landscape as a synthesis of seven concepts – regional, land-form, ecological, land use, heritage, scenery and parks and gardens. Naveh (1978:

57–63) defined landscape as 'the spatial and visual integration of the geo-sphere with the biosphere and man-made artefacts'.

Such quasi-scientific definitions of landscape do not provide us with adequate insights into the imagery that is so integral to resort-based tour-ism. The latter activity is, after all, dependent on the projection of highly selective images of landscape (beaches and the ubiquitous palm trees, for example) to *persuade* potential consumers that a particular destination is desirable. Since those doing the persuading have a specific behaviour-altering intent, an apparently objective interpretation of landscape will be inadequate. The subjectivity of landscape interpretation is emphasised in iconographic studies which seek 'to probe meaning in a work of art by setting it in its historical context and in particular to analyse the ideas implicated in its imagery' (Daniels and Cosgrove 1988: 2). It is this inter-disciplinary underpinning of the iconographic approach that is particularly useful for the present research.

RESORT-GOERS AS CONSUMERS OF LANDSCAPE

The iconographic interpretation of the South Pacific islands landscape highlights human interactions with resort landscapes as a symptom of modern life, an issue that Sack takes up in his analysis of postmodern 'places of consumption' (Sack 1992). Mitchell (1986: 2) has argued that images are signs which appear natural and transparent, but conceal 'an opaque, distorting, arbitrary mechanism of representation, a process of ideological mystification'. The views of Sack and Mitchell have challenged traditional interpretations of landscape. The closeness of the relationship between the promotional images projected by resorts and the behaviour of guests at these resorts does indeed seem to point to some of the contra-dictions and fragmentation of modern existence. In his essay *The Beholding Eye*, Meinig (1979: 33–34) has argued that there are ten possible perspec-tives which provide 'versions of the same scene' or ways of emphasising a view of landscape. Meinig's view implies that any ten individuals may see quite different things and place quite different emphases as they view a particular landscape: the 'eye sees what it wants to see'. The ten perspec-tives include nature, habitat, artefact, system, problem, wealth, ideology, history, place and aesthetic. Meinig has depicted the American view of landscape as having emphasised various different perspectives over time, most notably the issues of environmental awareness, image, symbol and representation, practical design possibilities and historical analysis. Such interpretations show the close association between the image and social dimensions examined in earlier chapters and the landscape focus of the present chapter.

Both the Fiji and Queensland islands are located in areas of outstanding natural beauty and, not surprisingly, both have been the subject of extensive environmental analysis. This is particularly the case with the Queensland islands. Their proximity to the Great Barrier Reef has led to a plethora of environmental studies by or on behalf of the GBRMPA. Despite the volume of work on environmental and landscape values, the existing research has been criticised by Craik (1991) as lacking any sense of social context. The present study therefore explores links between postmodernism and the landscape including the study of resorts as 'consumption places' where physical aspects are often driven by a commercial and marketing rationale. The approach taken includes the consideration of resorts as part of a global spatial hierarchy with the 'periphery' connected by certain commercial relations to the 'centre', usually the major metropolitan areas in the developed countries.

HISTORIC AND SPATIAL DIMENSIONS

As discussed in chapter 6, Bernard Smith has explained the nature of eighteenth- and nineteenth-century images of the Pacific (1950; 1992). The exotic flora, fauna and ethnography of the South Pacific raised some awkward questions about previously accepted 'global' principles and the region became a testing ground for the newly emerging scientifically-based world view. Smith has asserted that the emerging imagery of the eighteenth and nineteenth centuries and subsequent political and economic relations was an essential dimension of the imperial imperative: 'The European control of the world required a landscape practice that could first survey and describe, then evoke in new settlers an emotional engagement with the land that they had alienated from its Aboriginal occupants' (1989: ix). One might argue that current tourism imagery perpetuates such attitudes by 'taming' the landscape for the exclusive enjoyment of international visitors. In many cases, the appropriation may be perceptual rather than real. In Fiji, notably, the ownership of most (87 per cent) of the land by Fijians has certainly prevented the type of land alienation which took place in Australia as a result of the acceptance of the principle of *Terra Nullius*.

Landscapes are perceived differently by various types of visitor and by local residents. Bernard Smith has argued that those who live outside an environment are able to view a particular scene with detachment, whereas those who live there cannot. In the case of the Whitsunday Islands, the only permanent population (and many of the residents are not all that permanent) consists of resort employees and employers who have come from outside the area. In the Mamanucas, on the other hand, there is a small indigenous population, with its own perceptions of the land and by implication of the landscape.

The definition of the Fijian word *vanua* or land is symptomatic of the very different perceptions of land by locals (hosts) and incomes (guests). The word *vanua* is a powerful concept to the Fijian and is clearly differentiated from the *qele* (soil). The concept of *vanua* is inseparable from the people who originate from that land and from their socio-economic and political concerns. The words *levenivanua* (people) and *vanua* (land) can be used interchangeably with *levenivanua* meaning literally 'flesh of the land' (*leve* means flesh and, as previously mentioned *vanua* means land). According to Qalo 'the Fijians regard land as a part of their being. It is the centre of their lives and is a real source of security and purpose for living' (1984: 10). This view of the Fiji landscape is not (one might say cannot be) shared by Australian visitors. The attitude of Australian visitors will clearly be transitory and have been influenced by Fiji and/or Mamanuca images to which they have been exposed. Such imagery may even have prompted their selection of the Mamanucas as a holiday destination. As pointed out in chapter 7, the attitudes of guests and staff are linked in that, in 'service-oriented' business operations, staff will be expected to have an empathy with the views of guests, to ensure that they can anticipate their needs and provide responsive service. They are not 'culture brokers' in the sense used by Cohen to apply to tour guides (1972), but they may share the outlook of certain guests.

The study of resort landscapes has a pressing market rationale for island resort managers because emerging competition is eroding the monopoly that they once had over certain types of the 'exotic'. Why visit an island resort at great expense to experience palm trees, beaches and 'musak' when the same can be experienced close to home, often at much less expense? Urry also claims that 'spectacle and display' (often highlighted as island resort attractions) have become hallmarks throughout the world. Australian and Fiji resorts may learn from the experience of English seaside resorts which in the nineteenth and early twentieth centuries attracted most of the top entertainers of the period. The growth of the cultural industries in major cities gradually ended their relative monopoly and resulted in a growing perception that resorts were unexciting and offered little more than could be had close to home. Environmental contrasts are also being eroded.

Of the Whitsunday resorts, Hamilton has certainly been preoccupied in the past with attracting 'big name' entertainers. Since most Australians live at or near the sea, seaside holiday resorts have never offered such great contrasts with the home environment as in other parts of the world such as North America and in Europe. This means that island resorts promoting themselves in the Australian market have always been faced with competition from mainland beaches which enjoy mild climates and are usually located close to home. Nevertheless, Australia is experiencing a trend which Urry has observed in the UK, namely the increasing range of leisure facilities and options available in 'post-industrial' cities where much of the manufacturing industry base has disappeared. Settings such as Southbank in Brisbane with its artificial but atmospheric beach near the city centre, the

Southgate development in Melbourne and Darling Harbour in Sydney are all examples of the ready availability of outdoor leisure activities in the city. Whilst island resorts could previously appeal to an urge to 'get away from it all', the range of options available to satisfy such urges is increasing. Urry points to the attractions of the Centre Parcs development in Sherwood Forest, England with its double plastic dome, artificial seaside and permanent temperature of 84 degrees. 'In this complex, swimming is entertainment, fun and pleasure with tropical heat, warm water lagoons, palm trees and waterside cafes' (Urry 1990: 37).

Chapter 7 considered resorts as an element of postmodernism. To what extent do the physical properties of the resorts and of their natural surroundings affirm or challenge this view? One feature of postmodern consumption places is that they set themselves apart from the rest of the world (including the adjoining landscape), but at the same time they present themselves and those who frequent them as 'the centre of the universe'. One of the appeals of the South Pacific has been its position as an 'idyllic retreat' at the periphery (Sharrad 1990). As demonstrated by S. Britton, centre–periphery concepts still play a significant role in the structure of the international political economy. Urry (1990), however, argues that visitor attractions are becoming less place specific. A potent reminder of how the 'exclusivity' of attractions to particular places is being eroded is provided by the West Edmonton Shopping Mall in Canada. According to Urry, the Mall has turned the concept of geography on its head, where the periphery (Edmonton is situated in a fairly remote part of Canada) becomes the centre. Ominously for the South Pacific and Whitsunday resorts, one can stay in a 'Polynesian' room at Edmonton Mall and enjoy a reconstruction of the Great Barrier Reef.

Where such a state of affairs becomes prevalent, does one need to go to the places where these attractions originate? In fact displaced reconstructions can sometimes offer advantages over the original. Referring to the Great Barrier Reef Wonderland, in downtown Townsville, King and Hyde (1989a: 153) described the project as bringing 'the wonders of the Great Barrier Reef within reach of all visitors to North Queensland without weather, time or cost restrictions'. Taking such propositions to their extreme, one might argue that the modern technology of virtual reality could displace the need for tourism altogether. This book does not purport to investigate this scenario, but it is one that has leisure and resort planners around the world contemplating their future options. However, in terms of landscape, the threat to island resorts is that the essential physical elements of sun (solar devices), sea (water manipulated by techniques producing artificial waves) and sand (literally transplanted) can be reproduced in more easily accessible locations. One could argue that to justify the high prices that accompany island resort holidays, it will be necessary to provide more than the basic physical elements and to offer insights into the complex natural and socio-cultural phenomena that make up the destination. This is not to say that all holidaymakers are in search of 'authenticity', but it does point to a possible pre-emptive strategy.

POSTMODERNISM – ISLAND STYLE

Another central feature of postmodernism is the built environment which has been described as 'surely the sphere which . . . best demonstrates such a cultural paradigm' (Urry 1990: 120). There are many examples of this in the Whitsundays and the Mamanucas. The Fijian 'chiefly meeting house' at Sheraton Vomo Resort in the Mamanucas was specially constructed to house small conferences and meetings for (usually white) business/convention travellers and is an example of traditional functions confronting a modern function. The pastiche of styles so typical of postmodernism is also very evident on Hamilton Island where high-rise apartment blocks, Polynesian-style villas and a 'traditional nineteenth-century chapel' (largely used for Japanese weddings) are juxtaposed. Hamilton offers an eclectic range of accommodation and opportunities for avid shoppers. These examples typify what Berman (1982) has described as 'a unity of disunity', cutting 'across all boundaries of geography' and history, a 'perpetual disintegration and renewal, of struggle and contradiction, of ambiguity and anguish'. This is not to say that such postmodern concoctions will be unappealing, but it suggests that resort managers should have some awareness of the implications of such theoretical debate. This may be a tall order.

The architecture of the various Mamanuca and Whitsunday island resorts is diverse but can be generalised under what Joseph Emberton called the 'architecture of pleasure' (quoted in Parry 1983: 152–4). With this style 'everything was light, sun, fresh air and fun'. In a world where outdoor living is 'fashionable' such places 'enable consumers to express personal feelings and fantasies in a public language' (Sack 1992: 139). The common element is an attempt to create a setting for 'happiness'. This is similar to the ethos of theme parks, described by M. King as 'atmospheric parks – the happiest places on earth' (1981: 117). When we consider the aesthetics of the architecture in the two island groups, major differences appear. The vernacular bure-style settings of Fiji appeal to what Wright has called the 'abstract and artificial aestheticisation of the ordinary and the old' (1985: 230). The constructions are 'old' to the extent that most of the materials used are no longer typical of the residences used by Fijians which are nowadays constructed of cement, iron, fibreglass and other modern materials in preference to thatch and other more traditional elements. The atmosphere for Caucasian guests is one of nostalgia, fantasy and the exotic, nicely accommodating stereotypical images of the tropics.

Another feature of postmodern consumption places is their tendency to pursue social uniformity through excluding the type of 'undesirable elements' referred to in chapter 7, though it is true that this has always been the practice in elite tourism destinations. The 'exclusivity' label used so prominently in resort promotions has a number of spatial and locational dimensions. First is 'isolation', due to the higher cost of transport and related activities which restrict access to islands for lower income earners. As Hirsch (1978) has stated, the satisfaction that people gain from consumption depends on the consumption (or non-consumption) of others. Secondly islands are in short supply and are relatively small, leading guests to come into contact only with other guests from equivalent socio-economic groupings or 'lifestyles'. Whilst segregation cannot be guaranteed, cost and the style of marketing used can determine the overall social mix (see figure 8.1). Thirdly islands are separated from the backpacker clientele which congregates around Airlie Beach and Nadi. The most blatant social segregation in Fiji occurs on 'boutique' Turtle Island Resort in the Yasawa

---

H A Y M A N
I S L A N D

DELUXE LUNCHEON CRUISE

Departs 10:00am & Returns 3:00pm
$160 per person including lunch
Available Tuesdays & Thursdays

Join an exclusive group of up to 12 guests cruising with Circa Marine to elegant **HAYMAN ISLAND** where you will enjoy a sumptuous lunch served at the *"Beach Pavilion"*. After lunch you have the opportunity to explore the facilities in this magnificent 5-Star Resort, shop in the boutiques, stroll through the tropical gardens or take a swim in the resort pool.

Your luxury cruise includes:
* * return transfers aboard one of our luxury power cruisers including complimentary wine, beer and soft drinks*
* * lunch at HAYMAN ISLAND's "Beach Pavilion" restaurant (beverages not included)*
* * full use of beach and pool facilities at HAYMAN ISLAND*
* * stop off at Langford Reef for swimming, snorkelling and fish feeding*

Due to the exclusive nature of HAYMAN ISLAND, this cruise may not be available during periods of high occupancy at the resort, therefore **bookings must be made 24-hours in advance by phoning CIRCA MARINE on 58269 or ACTIVITIES on 58535**
(Circa Marine reserves the right to cancel due to weather or insufficient numbers)

Figure 8.1 Hayman Island: deluxe luncheon cruise
Hayman has sought to preserve its 'exclusivity' by limiting day-trippers to those Hamilton guests willing and able to pay the high prices.
Source: Barbara Harman, Public Relations and Marketing Manager (courtesy of Hayman Island)

Islands, where the owner/manager admits only guests who are judged able to contribute to dinner-table interactions. Children are not admitted. In practice, exclusivity has its limits since most resorts must attract a wide enough clientele to fill the rooms. Most Whitsunday island resorts welcome day-trippers and the promotion to this market does not emphasise exclusivity.

The promise offered by resort promotions to prospective consumers of a context with them at the 'centre of the world', is linked to the idea that he or she will be protected and set apart from incompatible social elements (see figure 8.2). Opponents of enclave-style tourism do not confine their criticisms to socially 'exclusive' properties but to any development which creates boundaries between holidaymakers and the surrounding natural and socio-cultural environments. The very word 'resort' has become associated with the concept of enclavism and segregation. Current market trends are moving away from this preference. Holidaymakers appear increasingly reluctant to cut themselves off from the immediate surrounding landscape (Poon 1989). The marketing literature which has identified such trends typically regards consumers as likely to bring about a better world through their rational and 'increasingly sophisticated' product choices. It is certainly true that changing consumer preferences have opened up a range of alternatives for resort managers. The danger is that the very resort managers who have the foresight to respond to such market trends may actually perpetuate practices of 'appropriating' the landscape by offering it for the (usually 'exclusive') enjoyment of their guests. Their resorts are typically described as 'your own private island', offering to take honeymooners to private and secluded beaches (H. Steinocker pers.comm.) and to offer 'exclusive' tours featuring the 'best' insights into the 'best' sites of interest. This type of 'appropriation' can lead to certain social costs where locals are excluded or offered only restricted access to such 'exclusive' places (though the only privately-owned beaches in either group are those on the freehold Malolo Lailai). Consumer pressure to create better insights into the local and regional area is not always an unqualified benefit for those who live there. Nor is the 'new' tourism regarded as 'better' tourism by all critics. Wheeler (1993) has coined ecotourists as 'egotourists'.

In island resorts we can see a tendency to merge public and private space, most noticeably through the privatisation of public space. Referring to shopping malls, Sack has stated that 'the front stages of malls and other places of consumption, like commodities themselves, combine elements from public objective space with elements from private, personal place. Malls want us to feel at home surrounded by friendly strangers. They are public places, privately owned' (Sack 1992: 146). Such places attempt to recreate the types of functions common to older city centres – public streets and paths, town squares and village fairs. The blend of public and private is also evident in resorts. Guests 'occupy' a particular unit or bure and make this their own territory. Many of the resort activities, however,

Figure 8.2 The cult of the individual at Club Med
Club Med's advertising promises consumers the opportunity to be at the centre of the world.
Source: Nigel Stoker, Managing Director, Club Med (courtesy of Club Med)

take place in areas which are markedly public in orientation. Nostalgia for the 'village of the mind' is often catered for by the layout of resorts.

Within the resort boundaries human behaviour is public but the rules and their enforcement are determined by private persons or entities (the owners and managers). The space immediately outside the resort boundaries is (usually) publicly or communally owned. The tendency of resorts to sell or promote the natural attractions of adjoining areas, means that the resulting 'way of seeing' such environments leads to a further blurring of the distinction between the public and private domains. The privatisation of public space is most blatant where local residents feel constrained from using beaches and other public areas because of perceived barriers such as concentrations of white sunbathers, though one might argue that such characteristics are less extreme than the practice of private beach ownership in countries such as Italy and the United States.

Another example of postmodernism blurring previously separate domains is its tendency to blend a nostalgia for an earlier and simpler era with a reassurance that modern conveniences and progress are never far away. Holidaymakers have differing levels of desire to experience the unfamiliar, but all demand some form of familiarity. As Sack (1992: 32) has stated, 'spatial routine can be liberating, allowing us to concentrate on other things'. While offering certain aspects that may be regarded as exotic or unusual, most resorts try to provide their guests with a sense of place. Activities such as welcome receptions and cues urging guests to think of their room as 'their own' assist in this process. Resorts attempt to provide guests with reassurance by naming their units with friendly-sounding titles such as the names of flowers and by using first names at the reception desk. In Fiji, the layout of bures attempts to strike a balance between privacy and communalism. The option of 'getting to know one's neighbours' is an activity allowing guests to determine where they would like to draw the line between the public and the private domain.

The other Whitsunday and Mamanuca resorts can learn one useful lesson from Hamilton Island. Though the pastiche of Hamilton is in many respects postmodern, Hamilton is associated with 'high-rise development' by consumers and by travel agents. The risk for Hamilton is summed up in Urry's (1990: 36) comment about the fate of the English holiday camps (of which most are now closed): 'the holiday camp was the symbol of the postwar society, reflecting the modernist architecture of the period'. It may be that Hamilton represents a style which is already *passé*. The other resorts can avoid the danger of 'overwhelming the landscape' as Hamilton appears to have done. It is notable that Hamilton is the only Queensland resort outside the main urban areas that has opted for a high-rise format.

RESORT LANDSCAPES AS CONSUMPTION PLACES

If Sack's (1992) principles, developed from theme parks and shopping malls, are indeed applicable to resorts generally, are they more readily applicable to either the Mamanucas or the Whitsundays? Alternatively, are they only applicable to selected and exceptional resorts? Do we see relative uniformity between the resorts in the two groups, or do they fall into distinct categories such as large and small?

A central proposition of Sack's work is the notion that landscape is offered to consumers as a commodity. He claims that such practices are typical of tourism and cites a publication entitled *Consumer's Guide to Florida* as an example of the commodification of (in this case) a whole state. Commodification is more pronounced still in the case of resorts. Anticipated tourist consumption plays a determining role in the layout of facilities which forecast the way in which landscape will be perceived. The resort is 'manufactured' for consumers in a way that even a tourism-oriented state like Florida, with its large resident population, could not be. Referring to department stores, Sack (1992: 135) states that 'commodities, stores and clusters of stores become landscapes that advertise both particular goods and consumption in general' and 'act as advertisements of the commodities that they sell'.

Sack has emphasised the role of places of consumption as an influence on our personal identity and sense of place. Landscape is viewed as a tangible representation of our concept of space. In a consumption place the landscape becomes a function of the way in which facilities oriented to selling commodities are dispersed and resorts 'commodify landscapes' and engender 'placelessness'. Like shopping malls and other 'consumption places', the resort landscape is devoted to consumption activities, with multiple commercial outlets to prompt consumer spending. Consumers are attracted to both types of place through advertising and associated promotional activity, thus imbuing the actual consumption which takes places at the destination with a series of connotations and preconceptions.

Klein (pers.comm.) acknowledged that 'cross-selling' takes place between the various food and beverage outlets at Hayman. Such practices are clearly prevalent in the larger resorts. It is questionable whether the smaller Mamanuca resorts incorporating only a single food/beverage and retail outlet are able to indulge in significant cross-selling. We may conclude that the theory is more applicable to certain resorts than to others. Resorts which have their own shopping malls are certainly examples of what Sack (1992: 144) describes as 'one commodity begets other commodities'. In the typology of the landscape of resorts developed in this research, the ones which fit the label 'consumption places' most neatly are those with multiple retail outlets. By contrast, the smaller island resorts are sanctuaries devoid of the billboards and other garish appeals to consume diverse products (including island resort holidays!) that line the main roads on the mainland.

Sack (1992: 2) believes that the physical layout of resorts is designed to encourage consumption and he states that 'a resort is not only a place in which things are consumed, but whose landscape is arranged to encourage consumption; and indeed, the appearance of that place – its landscape – is often the element that is consumed'. There are two dimensions to this argument. First is the *literal* consumption of goods such as food and beverages, retail items, sporting activities and entertainment for which a charge is made. Resorts have differing philosophies on whether payment should be made on a pro rata basis for all items consumed. Hamilton Island, for example, has traditionally attached a charge to most of the activities and services offered on-site. In contrast, the all-inclusive pre-payment approach practised by South Molle and Club Med makes it less advantageous for the resort operator to engage in 'hard selling' of resort activities since improved revenues are unlikely to eventuate. In reality all resort services are costed in some way, even if no guest payment is attached and the guest is unaware of the true cost of providing the service. In fact the distinction between immediate payment and pre- or subsequent payment may have become less relevant in view of the increasing use of credit card payments. In economic terms, the activity is still described as consumption whether payment is bundled or unbundled (for a discussion of this argument see Hooper 1994).

The various individual island resorts in the Mamanucas and Whitsundays exemplify different characteristics of Sack's definition of consumption places. Some of his principles will now be applied to specific island resorts within the two groups. Firstly, is it valid to argue that landscape itself is consumed? In the literal sense, islands which have transformed the landscape for a purely commercial activity have done so. Thus, parts of the landscapes of South Molle and Sheraton Vomo have been transformed into golf courses. One should acknowledge, however, that golf courses are atypical of the resorts that we are considering. Since suitable land is typically in short supply on islands, golf courses are more commonly found on 'mainland' resorts. Examples are Laguna Quays on the Whitsundays mainland and Denarau off the Nadi coast. Although all of the resorts in the Mamanucas and Whitsundays have areas specifically dedicated to commercial activity such as bars and restaurants, the emphasis on purchasing is less blatant than is the case with shopping malls where the exhortion to consume is ubiquitous. With resorts the main consumer decisions about 'consuming' the landscape are made prior to departure, when payment is made for the trip.

At Club Med an 'all-inclusive up-front cost leaves you free to choose what holiday you want to go on when you arrive'. The main component that Club Med does not include in the up-front cost is drinks. One might regard the use of 'beads' in place of money for the purchase of drinks as a clever ploy to give alcohol consumption an extra cachet and social acceptability. The practice is certainly not the antithesis of consumption! In the Mamanucas, the tariffs on Tavarua are 'all-inclusive' surfing packages. As explained in chapter 5, many other islands offer 'meal packages' which increase the proportion of total tourist expenditure prior to departure. In such resorts the landscape is not dominated by exhortions to spend, a fact reinforced by the consumer perception that neither the Whitsundays nor the Mamanucas were 'good for shopping'. In this sense the consumer experience is more akin to the practice of charging an up-front fee by theme parks as opposed to the more open-door free access offered by shopping malls and department stores.

The consumption ethic is blatant on Hamilton Island. The following letter to the *Times on Sunday* newspaper and quoted by Craik (1988) gives some insight:

> One is greeted with megabuck marinas, Onassis lookalike ships, an 18-storey high-rise obscenely towering into the sky and a convoy of mini-buses and battery-operated people-movers (at a charge) all provided to remind the visitors that this holiday means business – big business. If you happened to miss the point, a stroll through the Hollywood-set boutiques and eateries, or the sanctuary where wildlife has been nearly incarcerated, or a swim in the island bar set in a chlorinated pool, is sure to jolt you into an awareness that this holiday's level of adventure and activity will be highlighted by your endeavours to locate your wallet.
>
> (*Times on Sunday* 1987)

The more than seventy retail outlets on the islands are operated by concessionaires who compete with one another, selling to a captive market. This variety and relationship creates a consciously competitive environment akin to a market or shopping centre model. The very blatantly commercial layout of the resort landscape may have prompted the consumers (in the focus discussion groups) and travel agents (in the telephone survey) to the view that Hamilton is 'expensive'. The resort has also been the subject of criticisms by social scientists in the media (Sewell 1987). The apparent resistance from consumers and travel agents suggests that excessive and overt consumerism on island resorts may be poor business practice.

According to Sack, consumption places 'attempt to sever their connections to other process and places by presenting themselves as a world apart – a consumer's world, a showcase of goods and services, tours and vistas' (1992: 3). This interpretation applies to the more exclusive resorts such as Hayman and Sheraton Vomo. The former relies on the high consumer price to dissuade people travelling to the mainland (T. Klein pers.comm.). On the island, consumption is encouraged by creating the impression of a world of total exclusivity. The property promotes itself as a member of the exclusive club *Leading Hotels of the World*. It supplements such claims to exclusivity by association. It is 'closest to one of the leading attractions in the world [the Barrier Reef]'. Even the natural environment is being labelled as exclusive – an adjunct to the luxury of the resort itself. The perception of exclusivity created by such imagery is unquestionable. Ironically the philosophy now prevailing at Hayman is to create a more 'normal' and less 'precious' atmosphere. At Sheraton Vomo access to the island is by helicopter only. The reason given is one of ecology (boats will disturb the coral) and of providing a dramatic entry. It is also a commercial convenience that consumers will have little opportunity, nor probably any conscious desire, to move elsewhere from their cocoon of total exclusivity amongst their own type of people.

Sack's (1992) interpretation cannot be applied indiscriminately to all aspects of tourism in the Mamanuca and Whitsunday resorts. On Mamanuca islands such as Navini, Tavarua, Mana and Malolo, guests share the island with local residents (workers only in the two former cases). Certainly the living conditions of the two groups are different, with the locals occupying very basic accommodation. Nevertheless, the use of bure-style accommodation does gives guests a superficial impression that they are 'going native'. In the Whitsundays the process of resorts 'setting themselves apart' can certainly be sustained for Hamilton Island. However, the central role played by informal walks into the adjacent national parks on South Molle and Lindeman Islands are a type of linkage that limits the validity of Sack's argument. This is not to deny that national parks do possess cultural overtones which prompt a 'consumers' view' of their landscape. Nevertheless, one can argue that unlike consumption places such as shopping malls, department stores and theme parks, the Whitsunday island resorts do foster the opportunity for consumers to make connections with the adjoining and (relatively) non-commoditised landscape. The application of the 'world apart' idea is not absolute.

Another Sack (1992: 3) contention is more readily applicable to the Whitsundays and Mamanucas, when considered in a metaphorical sense: 'by severing our connections to the rest of the world, places of consumption encourage us to think of ourselves not as links in a chain but rather as the centre of the world'. All of the consumers interviewed for the present research had been exposed to promotional material for island resorts, even if a less formal medium such as word of mouth has played the major role in the decision-making process. It is the impact of promotion and (in particular) of advertising which combines the subjective ('lose yourself to find yourself' is a common theme in resort advertising) and the objective (for example, the price and facilities being offered). This practice of advertising to persuade consumers to view themselves as if at the centre of the universe has been commonly observed (Berger 1972) and undoubtedly applies to resorts including those in the Whitsundays and the Mamanucas.

Resort advertisements appeal to consumers by both offering an opportunity to define oneself and of setting oneself apart from others.

Sack argues that modernity has two interlocking dimensions which are clearly to be seen in his 'consumption places'. These dimensions are firstly 'optimistic and global' – a 'world virtually free of necessity, perhaps a consumers' paradise' and at the same time 'nostalgic and local' – longing for 'the virtues of the local community'. He regards consumption places which attempt to reconcile these conflicting pressures as symptoms of modernity, typically 'territorial communities of convenience, which lack true purpose and cohesion' (1992: 7). The resorts of the Whitsundays and Mamanucas certainly exhibit these characteristics to varying but generally substantial degrees. The bure-style accommodation of the Mamanucas and the village-style settings with centrally-located public areas where guests can mingle, are certainly nostalgic and local in their appeal. At the same time the liberal use of labels such as 'paradise', 'abundance', 'pampered' and 'luxurious' in the promotion of both groupings is clear evidence of 'modernity as optimistic and global' in action. In the Whitsundays, the nostalgic and local is less in evidence at a number of the resorts, but is still there. It is obvious at the smaller resorts of Hook Island and at Palm Bay. It is also present at the Club Med Lindeman Island 'Village'. The use of the name village owes more to the historical development of the simple Club Med formula but still implies an intent to create a 'village-like' atmosphere. At South Molle, Brampton, Hayman and Daydream Islands the village setting is less pronounced and the fact that all are operated by large corporations introduces a 'corporate' dimension to the pervading atmosphere, albeit a generally subtle one. All resorts have public communal facilities, though layout, scale and design all influence the extent to which a 'village' feel is engendered.

The response to the urge for the 'nostalgic and local' is catered for by the promotion of swaying palms and tropical tranquillity. The importance of escaping from the 'rat race' was expressed by a number of consumer focus group participants and the security offered by the islands for children was contrasted in particular with the dangers present in the suburbs of major cities. The security dimension was not associated with the presence of security systems (though it was mentioned that such systems would be necessary in destinations such as Bali), but rather by their absence. Safety was seen as a 'natural state' in Fiji island resorts. The most brazenly 'modern' of the resorts is Hamilton with its urban form and vehicular transport. Hamilton claims to have something for everyone with its 'Polynesian-style bures' constituting a traditional, albeit contrived form of accommodation with particular appeal for honeymooners (B. Diamond pers.comm.). Generally the mixture of cosmopolitan and local/nostalgic is evident in all resorts, though with a greater emphasis on the latter in Fiji with its freestanding units. One may speculate that the less nostalgic and more 'corporate' approach of the Whitsunday resorts may have prompted the consumer view, expressed in some focus groups, that the Whitsundays are 'less friendly' than overseas destinations such as Fiji. Whilst other factors are at work, the view that 'Australians are less friendly when holidaying in their own country' is more likely to be the result of the atmosphere or layout of resorts, rather than any fundamental truth concerning behaviour at home and overseas.

In contrast to shopping malls and theme parks, resorts are typically integrated into the immediate environment. Whilst practice often falls short of this ideal, the resort commonly seeks to attract visitors on the basis of the pristine environment offered nearby, plus the quality of facilities and activities *in situ*. The resort is selling the destination first and the facilities second. The landscape and the resort itself make up a composite product.

Some of the larger island resorts have more in common with cities or towns than villages because of their consumption orientation in the form of activity centres, boutiques, restaurants, golf courses and bars. At Hamilton Island, the intense competition between the various restaurant and other lessees, results in a 'harder sell' approach conveying a strong sense of urbanisation. Most resorts in the two groups, however, choose to portray a 'village-style' atmosphere. Resorts might also be regarded as providing an urban 'shop window' into the otherwise rural setting of the destination in which they are located.

It is interesting to consider the various resorts using Lane's concept (1994) of urban and rural tourism, with the two at opposing ends of the continuum. Lane's analysis underpinned the recently released *National Rural Tourism Strategy* (Commonwealth Department of Tourism 1994b). Physically, Hamilton is the most urban, though it 'sells' the natural (and rural) beauty of its surroundings. The smaller Mamanuca resorts with the closest community integration are the most rural.

A SYSTEMS APPROACH

Sack's and Urry's theoretical propositions help us to explain the meaning of the relationship between resorts and their immediate environs. They are less useful as practical guides to the optimum integration of landscape concepts which enhance the appeal of resorts and their sustainability. In her writings on resort landscapes Hana Ayala, a Professor of Social Ecology at the University of California, has adopted a more pragmatic, management-oriented approach which gives greater emphasis to market trends, consumer expectations and the extent to which responsible landscape concepts can be made marketable (Ayala 1991a; 1991b). She has urged the need for resorts to broaden their concept of product and marketing by taking into account the landscape in which they are located. She has noted a global trend towards resorts occupying a larger share of the total number of hotel rooms (16 per cent of all hotel properties by 1988) and a trend towards bigger individual properties (1991b: 280). She claims that resorts can enhance their competitive position by developing a sense of 'placeness'. She has also argued that 'resort developments tailored to tourists of a specific nationality will probably experience the kind of "crisis of dependence" that is currently eroding the tourism industry in Malta' (1991b: 281). She seems to argue that too much pandering to the tastes of a particular market will reduce the longevity of the resort's appeal, compared to the approach of developing a close association with the destination environment. The present research has observed that the Mamanucas have been tardy in reducing their dependence on the Australian family market and the Queensland island of Great Keppell probably adhered to its 'youths only' policy for too long.

Ayala argues that market trends (or 'megatrends' as she describes them) indicate consumer demand for resorts which emphasise environment. She depicts tourism as increasingly 'sightseeing oriented, with tourists searching for new experiences' (1991b: 281). Whilst concentrating on the creative possibilities of incorporating the environment, Ayala also acknowledges the dangers inherent in transnational corporations attempting to impose a 'standardised' product in diverse environments. She gives the example of the recently renovated and extended Hyatt Regency Saipan which according to a recent press release will use the 'water and foliage theme for which Hyatt Hotels are world famous. Fountains, waterfalls, ornamental lakes, pools, bridges and an abundance of foliage and flowering plants will create an idyllic setting for this romantic resort oasis' (quoted in Ayala 1991a: 583). There is clearly a fine line between adventurous landscaping (for which Hyatt is famed) and creating a sense of placelessness because a certain type of 'exotic' environment is transportable to anywhere in the world. The dangers of aping stereotypical images of the exotic are particularly tempting for South Pacific destinations such as Fiji because so little adaption from actuality to image seems to be required. In practice the Whitsunday resorts of Hayman and Daydream are probably closest to the type of concept applied in Saipan, with their impressive, but ultimately intrusive architectural designs.

Five of Ayala's hotel landscape themes offer useful perspectives for the Whitsundays and the Mamanucas. She refers to 'processing the ecological legacy of the site' and several examples of this are evident in the two island groups. The creation of the marine reserve adjacent to Beachcomber and Treasure Islands is one example, though overall 'processing' is undertaken in a less sophisticated way in the Mamanucas than in the Whitsundays. This applies to both environmental management and to interpretative material. In the Whitsundays, GBRMPA ensures that management procedures are formalised and in place and that the presentation of interpretative material is developed.

221

Her concept of 'fusing interior and exterior' can be readily applied in island resorts because 'geographical or ecological boundaries rather than the property lines, define the landscape system for the guest' (Ayala 1991a: 577). The graded paths and tracks winding directly from South Molle and Club Med to the national parks act as a lure to houseguests and provide a fusion of site and setting. In this context, the island resort label is an asset and a potential curse. The archetype of the deserted tropical paradise is an appealing promotional message that tempts resort operators to use metonyms in their promotions, implying that the resort *is* the island. Where metonyms lapse into tropical island stereotypes they lose their distinctiveness and appeal.

'Place-identity through contrasting landscape experiences' (as described by Ayala) is evident at Mana Island Resort with its Fijian village on the island helping to evoke this contrast. Overall the 'packaging' of landscape experiences had been less formalised in the Mamanucas. For the most part, landscape has been sold as a 'setting' for other activities and not as the primary drawcard. None of the Mamanuca resorts, for example, participate in the Select Hotels and Resorts International's *Select the Natural Pacific*. This programme enables the resort guest to 'experience the essence of local ecosystems through intimate contact with nature' (Select Hotels and Resorts International 1993: 1). It demonstrates the range of alternatives open to Pacific resorts which wish to integrate fully the resort experience with intimate environmental encounters. The Select Hotels initiative is significant since it attempts to blend 'conventional' resort accommodation with activities typically regarded as specialist ecotourism. In the Whitsundays, contrasting landscape experiences have played a small role in promotional material but are significant as activities at the resorts. Examples include island walks, reef trips and visits to natural settings such as Nara Inlet (with its Aboriginal cave paintings) and Whitehaven Beach.

'Landscape tailored to evoke local flavour' is evident at Hayman Island Resort where the developers brought in botanical species from throughout Australia and developed the resort gardens into an attraction. Management at Hayman also introduced Australian fauna onto the island. Historically the Whitsundays have been guilty of 'appropriating' Pacific and particularly Polynesian imagery which may actually detract from the genuinely local Whitsunday flavour. There is a long tradition of staging Polynesian floorshows at livelier resorts (such as Royal Hayman, predecessor to the existing development). Palm Bay Hideaway accommodates guests in 'Polynesian-style' bures and (as previously stated) Hamilton Island in 'Polynesian-style units'. To date, Queensland colonial-style architecture has helped provide mainland Queensland resorts with a local flavour (the Hyatt properties at Coolum and at Sanctuary Cove are examples), but this challenge has not yet been taken up in the Whitsundays. It is an issue which deserves to be taken seriously by the WVB and it is notable that the Draft Whitsunday Tourism

Strategy does flag its intention to pursue a regional style in future resort developments (Office of the Co-ordinator General 1994).

Another expression used by Ayala (1991a) is 'designing greenhouses, islands and oases'. At Brampton Island, the saltwater pool perched on rocks above the ocean can be seen as an enhancement of the resort's seashore relationship and as compensating for the separation of the resort from the beach. The rocky setting of the Brampton pool is like a continuation of the shoreline. Its elevated setting maximises the coastal panorama. Its use of salt water gives an extra link with the larger environment as setting for a greater drama. Ayala talks about the use of islands as design formulae themselves. She says that often 'islets are set in pools and lagoons to endow the water with tropical and exotic moods' (1991a: 582).

Ayala's North American perspective can be made more relevant to Australian conditions through applying the concept of *curtilage*. This term is derived from Middle English and has been defined as the 'area attached to a dwelling-house as part of its enclosure' (Australian Concise Oxford Dictionary 1987), but has become 'Australianised' through its use in the Burra Charter, a document which established heritage responsibilities under the International Convention on Monuments and Sites (ICOMOS). The majority of self-contained resorts do not conform to traditional concepts of heritage which have been associated with curtilage. Typically, though, they claim a close association with a particular natural heritage and thereby enhance their status and significance as a destination. Curtilage involves the recognition that a particular site of significance can best be understood as occupying the centre of a series of concentric rings. The different areas accounted for by the different rings will have different values attached to them. In the case of a resort the immediate area might consist of the nucleus of resort buildings, followed by the site (incorporating say, golf course and pools) followed by the island itself (including beaches) and finally the region. Because of the resort's dependence on landscape appeal to achieve saleability and therefore market value, it seeks to create a value upon the linkages. The easiest way to create this association is through promotion. To provide funds or assistance to protect adjoining areas is a different matter, since such investments may not guarantee exclusivity of use for houseguests. In such a 'common pool' arrangement, there is little incentive for a private developer to invest in something that may benefit a competitor. Resort managers need a mix of incentive and external pressure to preserve adjoining environments. Most acknowledge the importance of a pristine environment to the appeal of a resort (King and Weaver 1993), but few are likely to take responsibility unless they see competitors making equivalent 'sacrifices'.

As frequently occurs in the assessment of curtilage for places of heritage significance, resorts have an ambiguous relationship with the environment outside the physical boundaries of the property. One can argue that a central business district hotel (for example, the Grand Hyatt in Melbourne) bears no responsibility if an environmental problem occurs nearby (for example, if heavy rainfall causes the beaches of Port Phillip Bay to become unsafe for swimming). If the attractiveness of Melbourne as a destination is affected due to associated negative publicity, some visitors may choose to travel elsewhere and Hyatt may lose some business. The Hyatt may, however, legitimately claim that it has no control over the offending events. Some city-based properties are located adjacent to public space (the aptly named Melbourne Hilton on the Park is an example), but most abut private space such as office blocks, thereby minimising the influence of management over the activities that take place there. The external environment is less integral for their appeal to consumers, relative to location and quality of facilities.

In contrast, if an island resort area such as the Whitsundays were to be affected by pollution (for example, an oil spill), the resorts themselves would be perceived as accountable. The natural environment is integral to the resort experience. Because of the interdependence of the two elements, it is often considered important that the visual appearance of the resort complements the adjacent environment. South Molle, Brampton and Club Med Lindeman Island are all surrounded by national parks. This proximity blurs the distinction between the man-made and the natural environments into a seamless web. It is significant that the Tourism Development Programme for Fiji (UNDP 1973) stressed the need for resorts which were low-rise and in thatched form. This form was regarded as complementary to the physical and social environment of Fiji.

In the Mamanucas, Plantation Island is the only resort offering 'motel units' as a (cheaper) alternative to bure accommodation. Beachcomber features some dormitory accommodation and Mana Island has some 'duplex units'. Finally Musket Cove has a small number of 'villas' as well as some bure-style timeshare accommodation. Despite these exceptions, the unity of form amongst the island group is striking. All developments are well concealed by vegetation. No resort attempts to reproduce an urban environment. An obvious contrast with the Whitsundays is one of scale. The two largest developments in the Mamanucas are Mana Island (with 132 units) and Plantation Island (110 units); both devlopments are well concealed behind vegetation. In the case of Mana, the resort is particularly dispersed, though this may change with the proposed expansion of the resort (K. Palise pers.comm.). Were the plans to be put into effect, the first large resort would be introduced into the Mamanucas. For the moment, the resorts are all relatively small and quite well concealed.

The Sheraton of Fiji and Regent of Fiji are large resorts located just outside the Mamanucas. Both are managed by international chains and are on a comparable scale to the Whitsunday resorts. The Sheraton is blatantly 'international' in style, whilst the Regent makes more concessions to a

Fijian style with the mood of a traditional village set amidst tropical gardens. It is noticeable that the selection of the reclaimed Denarau Island as the site for these resorts, which make up Fiji's major tourism development, may have been influenced by its potential to remain relatively free of local politics (though the operators do pay lease monies to the landowners through the NLTB). EIE, the Japanese owners, claim that Denarau is the largest integrated resort in the Pacific islands, an indication of its dedication to consumption. All areas of the island are revenue generating, ranging from the marina to the golf course and sporting centre. A detailed study of Denarau will not be undertaken in the present research since it lies outside the Mamanucas. Its presence should not be ignored, however, since the incorporation of Denarau into the Mamanucas region would potentially enable the area to undertake a major regional marketing repositioning.

In the Whitsundays the obvious exception to the practice of blending physical resort development with the adjacent environment is Hamilton Island, where the high-rise configuration has often been described as 'Surfers [Paradise] on an Island'. Hamilton Towers can be seen for many miles around, making it a prominent feature of the Whitsunday landscape. Though it is not visible from the other Whitsunday resorts, it quickly comes into view as one's inter-island launch sets out from most of the other properties. At the other extreme is Palm Bay Hideaway which consists of fourteen cabins and bures adjacent to the beach. The small scale of the development, the abundance of natural camouflage (palm trees) and use of individual, freestanding units gives the 'resort' a low-key appearance. In the case of the larger Whitsunday resorts, the scale of development does not allow for a predominance of freestanding units. Instead, construction is often in the form of multiple 'wings' (for example, at Hayman) with a central foyer (as on Daydream). The view expressed by the WVB (D. Hutchen pers.comm.) that the Whitsundays offer 'something for everyone' is more rhetoric than fact. The tiny Palm Bay Hideaway is the only property of its type, and the style of accommodation that it offers in that particular tranquil setting is in short supply relative to the total resort capacity in the Whitsundays.

Whether the style of construction is considered effective is substantially a value judgement. The General Manager of Daydream (B.Vincent pers. comm) described the developers of that resort (Jennings Industries) as having deployed the most up-to-date construction techniques to create a sympathy with the environment. Certainly the natural environment is re-created throughout the resort with lush, well-watered vegetation. This extends into the cavernous foyer area of the resort and helps distinguish it from equivalent foyer areas in city hotels. Nevertheless, the author noted parallels with the foyer of the Capital Parkroyal in Canberra (also operated by the Southern Pacific Hotels Corporation (SPHC)).

The styles of Hayman and Daydream Resorts are grandiose and monumental. Both incorporate enormous swimming pools which act as centre-pieces for the resorts (and are often highlighted in promotional material). Both have huge cathedral-like foyers. In the case of Daydream, the various resort wings run into the central foyer area. In Hayman, each wing has its own foyer. Presumably this is an expression that its huge development budget would allow Hayman to 'rise above' the stereotypical resort construction.

Another common feature of Hayman and Daydream is the use of a tropical style of vegetation. The contrast between the island environment (dry and sparsely vegetated in the case of Hayman) and the actual resort environment (well watered, lush and tropical) is more striking than in the case of Daydream. In the case of the latter the tiny dimensions of the island have meant that the resort has all but *become* the island. The lushness looks less incongruous than at Hayman but this is perhaps symbolic of the dominance of man over nature and of the artificial over the natural.

The Whitsunday resorts do not conform to an obvious resort environment 'formula' and defy simplistic type-casting. South Molle Island gives the arriving guest the impression of the traditional English nineteenth-century resort with its extended pier, promenade along the edge of the water and construction parallel to the waterfront. Club Med Lindeman island is built on a steep hill and owes more to the Italian resorts of the Amalfi Coast with their panoramic outlooks to the sea beyond. Brampton complements the natural environment with its double-storey Queensland villas. It is at some distance from arriving watercraft (guests are transferred via the island train or from the airport). Radisson Long Island Resort, with its relatively small scale, is fairly well concealed by the natural environment. It is like a traditional housing development with extended two-storey white-washed blocks. Hook Island Wilderness Lodge is on a smaller scale still and is concealed in a small cove. The atmosphere is informal, with the pre-dominance of campers over those occupying the formal accommodation.

CONSUMER AND INDUSTRY PERSPECTIVES

The perception of resort managers towards issues of environmental management issues is predictable. Managers embrace environment management techniques which promise to reduce costs (the installation of shower roses to reduce water consumption, for example). Second, the fact that the major tourism attractions of the Mamanucas have been 'sun, sea and sand' leads to the response that 'it's in our own interests to look after our environment since that is what attracts the tourists in the first place'. Other issues are less predictable, for example views on the relative responsibilities of resorts and government towards environmental protection. The author asked the managers whether their Australian clients have an interest in environmental issues and in learning about the cultural heritage of the region and if they believe that such interest exists, what initiatives they have taken.

Some Mamanuca islands have Fijian villages and no resorts whilst two (Malolo Island and Mana Island) have resorts and villages. The degree of interaction between guests and the local community is well developed in all resorts, including the links between resorts located on previously uninhabited islands and the relevant landowning community. Jim Saukuru stated that 'the fact that the resorts are so interwoven with the villages and that they are so close to us is a real advantage. Guests really appreciate being able to visit the Fijian villages in the Mamanucas – it gives them a real taste of the culture and lifestyle' (pers.comm.).

One resort which is approaching the 'natural and cultural heritage immersion issue seriously' is Naitasi (M. MacDonald pers.comm.). It is unique in the Mamanucas in being a small resort (thirty-eight units) on a (relatively) large island with two separate villages. This pre-existing Fijian 'infrastructure' provides the resort with a greater opportunity for cultural and environmental differentiation. The manager has plans to pursue 'the ecotourism side of Malolo . . . to enable diversification beyond sun, sand and sea, though you can never entirely escape from these key elements'. Guests will explore 'grave sites, shell middens and pottery shards, rock structures with burns on the roof that were old fireplaces'. Six village guides are being trained at the time of writing and a *Walk Malolo* leaflet is being prepared in conjunction with the Fiji Department of Tourism. This leaflet will also include details of 'garden walks' which will enable tourists to identify the island flora and fauna and help to provide traditional beach-goers with a 'total package'. In conversation MacDonald and his wife Sandy admitted that there was a general lack of awareness of the significance of cultural activities amongst the villagers, particularly concerning historic or heritage sites. Nevertheless the Naitasi initiative is a clear attempt to blend classic sun, sea and sand markets with cultural and environmental diversification. Naitasi, like other resorts, takes guests to Yanuya island in the Mamanuca-i-Cake group. On Yanuya is one of only four villages in Fiji where traditional pottery is made. The historic dimension is highlighted by the fact that one 'can pick up pottery shards on the beach that are dated 2,800 years old' (M. MacDonald pers.comm.).

As in the Whitsundays, the Mamanuca managers expressed surprise at questions about environmental standards, stating that resort owners and managers would obviously protect the environment, since that is what attracts visitors to the islands. Shaw's response was typical (pers.comm.): 'Why kill the golden goose? Basically we are here to make a profit and our visitors expect sandy beaches, clean water and no litter or rubbish.' Beyond such superficial responses, the standards prevalent in the Mamanucas are clearly less consistent than in the Whitsundays. Many managers commented on the advanced state of environmental management standards in their own resort, but talked disparagingly about practices on certain other islands. In contrast, the Whitsunday managers stated that standards were very high throughout the group (though a number did claim that their own facility had the 'biggest and the best' treatment plants!). The sewage treatment on Hamilton is certainly well concealed and it is the smaller resorts where problems exist. At Hook Island, Tall claimed that the impending regulations requiring an upgrade of sewage treatment was an unnecessary and unwelcome imposition for his resort (pers.comm.). The system of sewage and rubbish disposal operated at Palm Bay was notable for the periodic unpleasant odours that emanated from it, close to the accommodation area. The latter two instances highlight the environmental dilemma for smaller resorts where the minimum legally enforceable standards are becoming increasingly stringent.

The Mamanuca managers acknowledged some less desirable practices, though within the context that 'standards are constantly improving'. According to Smith (pers.comm.), 'Castaway put in a sewage treatment plant and we put in a similar one. Beachcomber and Treasure pump their effluent (not raw sewage) a mile or two out to sea. It doesn't seem to have had any effect, in fact the fishing has been good! Plantation Resort is planning to put theirs out to sea, but I would resist that because we are too close to the reef. At Castaway, Plantation and Musket Cove we used to dump our garbage in the sea – now it's all sent back to the mainland or burned and buried on the island – on this one we have enough space for that' (pers.comm.).

On Treasure Island 'all refuse is bagged or frozen and sent back to the mainland. At this resort no refuse is thrown into the ocean. We have to ferry it away because we are so small' (J. Saukuru pers.comm.). Pride in the removal of waste to the 'mainland' is of course no assurance of proper disposal, especially in light of the inadequacy of the Nadi-based disposal system highlighted in the *State of the Environment Report* (IUCN 1992).

Whilst the intent may be sound, it is clear that the smaller Mamanuca resorts lag behind the larger Whitsunday ones in their approach to waste management. The environmental edge held by larger resorts was made forcibly to the author by the Head of EIE Fiji, owners of the Denarau Resort (A. Thompson pers.comm.).

The General Managers at both Sheraton Vomo and Daydream Island stressed the requirement that recently completed developments (redevelopment in the case of Daydream) must pay particular attention to environmental requirements and make maximum use of the available technology (pers.comms.). Vomo uses 'the most modern sewage treatment available. We have our own desalinator and our own sewage treatment plant with run-off through a reticulated irrigation scheme. Our garbage is sent to the mainland for compaction and where possible our cans are recycled'. The Sheraton PR machine was into full swing when the golf course layout was discussed: 'Only three trees were moved to build our nine-hole golf course. It was designed around the trees' (N. Palmer pers.comm.).

In the Whitsundays, the legislative and regulatory requirements associated with marine and national parks result in stricter restrictions than the unsystematic approach in the Mamanucas. The larger Whitsunday resorts also benefit from the economies of scale required to justify certain types of capital-intensive plant. The increasing practice of Whitsunday resorts to proclaim their environmental management credentials, particularly in the cases of Hayman and Daydream, is also notable. Public relations on behalf of these resorts has pointed to their winning awards for environmental management practice. The Mamanuca resorts have not attempted to raise awareness of their environmental achievements and risk appearing as laggards relative to competitors in areas such as the Whitsundays.

In the Mamanucas the move towards environmental responsibility has been prompted by tourism marketing considerations (what will our guests be willing to accept?) more than by regulatory dictate. The positive dimension of this is that the resort managers seem fairly aware of customer expectations and conscious of the close relationship between the quality of the environment – particularly the marine environment – and visitor satisfaction. The negative dimension is that in the absence of an established, properly monitored regulatory scheme, the environmental well-being of the area is dependent on responsible behaviour by all operators. In the absence of full consensus amongst the various members of the Mamanucas branch of the FHA (as evidenced by the total lack of co-operation on certain issues), this is a risky assumption.

There are no incentives in place to encourage environmental responsibility by the Mamanuca resorts. The types of environmental awards made in Australia are not practised in Fiji; there was one environment category in the 1991 Fiji Tourism Awards, but the awards scheme has been discontinued. Such awards help to focus attention on environmental 'best practice' and from the discussions in the Mamanucas, it is clear that such an educative process is urgently needed.

In a recent press report, FVB Marketing Services Manager, Ilisoni Vuidreketi stated that 'What the visitor of the mid-1990s wants is culture, history and an outdoors environment (with perhaps a bit of sun, sea and sand on the side). Fiji will be the world's first ecotourism destination. And that . . . is official' (*Asian Hotelier* 1994: 10). Two dangers are evident for the Mamanucas. First, expectations of environmental standards may not be fulfilled if too many promises are made before controls are put in place. Second, the rhetoric that sun, sea and sand are a case of 'perhaps a bit . . . on the side' appears to devalue the significance of what has been, to date, Fiji's main tourism market. Such views reinforce the perception that recent Fiji promotional campaigns of the 'Discover the Fiji You Don't Know' type were already trying to promote tourism activity away from the established areas such as the Mamanucas, and were almost hostile to the whole concept of 'sun, sea and sand' (M. MacDonald pers.comm.).

Within the Mamanucas, environmental visitor interpretation is less developed than in the Whitsundays where island walks are led by National Parks and Wildlife Service (NPWS) guides (on Lindeman Island) or by guides trained by NPWS staff (on South Molle Island). In the Mamanucas, interpretation is constrained by the lack of identification markers and by the absence of leaflets and other materials advising visitors about the adjacent attractions. This issue is certainly being addressed by Naitasi (J. Francis, pers.comm.), but other islands are lagging (J. Francis pers.comm.). Day tours of Malolo Island are advertised in the daily *What's On* at Musket Cove which is a positive initiative, though a major constraint is one of providing quality interpretative materials to ensure a 'quality' experience for participants. The opening of Sheraton Vomo might provide a stimulus to other resorts to introduce quality walking tours and interpretation, given its greater financial flexibility to indulge the guest's every whim. It does list amongst its 'things to do' section, 'flora and fauna island tours' and 'Vomo Island discovery tours', and certainly has the potential to be packaged in a way which emphasises the environment (as other Fiji resorts are in Select's *The Natural Pacific*). Current brochure material takes a different approach, divided into the following sections: 'Vomo Luxury, Vomo Cuisine, Vomo Service and Vomo Leisure'. Environmental and cultural immersion appear to be absorbed within 'Vomo Leisure'.

At this stage in their development prior to the implementation of management structures, it would be unwise for the Mamanucas to place much emphasis on their environmental credentials. Consumers are cynical about exaggerated use of the label 'environmentally friendly' and would be quick to identify defects. General consumer cynicism about the marketing of 'environmental friendliness', combined with the inadequacy of current environmental protections, suggests that an emphasis on structures should precede promotional activity. The same applies to land-based environmental tours which are another area of current enthusiasm and interest. Care is needed to ensure that the range of quality tours with genuine environmental safeguards are in place, before Fiji claims the area inland from the Mamanucas as a world-class ecotourism destination. The consumer focus groups did not think that the Mamanucas were a suitable destination for those 'who like to get out into the bush' and the Whitsundays were regarded as currently more suited to such activity. This low awareness of the suitability of the islands for bush-walking indicates that any marketing focusing on this type of activity should be low-key in the first instance.

All twenty-one resorts examined in the present study involve an intimate and intricate relationship with the surrounding natural environment. Management at all resorts acknowledge the key role played by natural surroundings as a means of attracting visitors. Both areas offer settings which adhere to the traditional romantic European 'tourist gaze' – clear water, mountainous or at least undulating islands, coral reefs, sandy beaches and sunny weather. Both are endowed with 'competitive advantage' when assessed in terms of their natural settings. The level of intimacy with the natural environment offered to visitors is, however, quite diverse.

The Mamanucas resorts have followed an overall development concept (UNDP 1973) which emphasised low-rise development using traditional style accommodation. The Whitsundays have adhered less strictly to a total concept, resulting in a diversity of styles with certain developments succeeding better from an architectural point of view than others. Whilst the eclectic approach has its supporters, it has impeded the emergence of a strong regional concept. The fact that a single development (Hamilton Island) has been able to sour consumer and travel agent perceptions points to the dangers of the eclectic approach. The prevalence of motel-style accommodation also limits the intimacy with the environment offered by the experience.

Sack's (1992) theory is most clearly articulated by Hamilton Island and less so on the smaller resorts where there is less blatant merchandising activity. Whilst the chorus of criticism of 'appropriation' of the landscape will increase, as Ayala's (1991a; 1991b) writings have shown, the integration of landscape concepts can be seen as a positive development. Consumers benefit from more rewarding holiday experiences, resort managers benefit because such concepts allow the resorts to differentiate themselves from current and potential competition and the environment benefits because resorts improve their environmental management practices.

Resorts can usefully adopt more environmentally inclusive policies in fairly developed tourism regions such as the Whitsundays and the Mamanucas. Such approaches do not preclude the channelling of holidaymakers into more remote and less developed areas (as is being attempted in Fiji). The Whitsundays and the Mamanucas will not become archetypal 'ecotourism' regions, but both can benefit from a more inclusive approach to landscape. Such an approach will not end the tendency of resorts to become 'consumption places', though it is likely that it would preclude further developments on the scale of Hamilton Island. Consumer and travel agent responses have clearly shown that developments on this scale are seen as overwhelming the landscape and not as complementing it.

9

OPPORTUNITIES FOR FURTHER RESEARCH

Resorts provide a setting for human activity which has far-reaching social and environmental repercussions and should not be dismissed simply as hedonism. They have often been disregarded as trivial and unworthy of serious scholarly discussion, though the recent writings of MacCannell (1992), Sack (1992) and Urry (1990) indicate that some scholars are taking this heightened form of tourist place-making very seriously indeed.

A key element of comparative studies is the highlighting of similarities and contrasts. Some similarities are geographically specific and of limited global validity. Other similarities, such as the functioning of islands as consumption places, have a universal validity and can help in the development of a comprehensive theoretical basis for the study of island resorts elsewhere. Some key differences and similarities are summarised in tables 9.1 and 9.2.

Table 9.1 Differences between the Mamanucas and the Whitsundays

Mamanucas	*Whitsundays*
International destination for Australians	Domestic destination for Australians
Low awareness of the region in Australia	High awareness of the region in Australia
Small to medium sized resorts only	Small, medium and large sized resorts
Many tour operators featuring the islands	Few tour operators – airlines dominant
Part of a developing country	Part of a developed country
Marketing emphasis on local culture, setting and relaxation	Emphasis on natural beauty and relaxation
Travel agents officially paid 9–10 per cent commission for air tickets	Travel agents officially paid 5 per cent commission for air tickets
Insignificant presence of branded hotel chains	A number of chains present (e.g. SPHC/ Holiday Inn)

Source: Author's interpretation of survey results

232

Table 9.2 Similarities between the Mamanucas and the Whitsundays

---

Both regions are characterised by a tropical setting, pristine beaches and coral

Heavy dependence on a small number of airlines serving the Australian market

Easy accessibility of a major airport

Short-haul destinations

Direct access available via both watercraft and aircraft

Both are established tourism regions

Co-operative marketing undertaken by the various resorts in the groups

Both have fabricated tourism island names (e.g. Castaway, Beachcomber, Daydream) plus original names (e.g. South Molle, Tavarua)

Source: Author's interpretation of survey results

It has been demonstrated that the Whitsundays have advanced significantly towards an integrated regional tourism concept, with fairly stable boundaries. All of the Whitsunday resorts appear committed to the regional concept. The Whitsundays enjoy a solid and strengthening regional structure backed strongly by resources from the relevant state tourism commission. The region's marketing activity has been successful in the recent period, providing an impetus for the regional tourism bureau to extend its involvement into product development and quality assurance. While competing interests are still in evidence, the complementarity of the well-developed island resorts and the smaller mainland properties has emerged as a key strength. At last enjoying high occupancies, the Whitsunday Islands with their constrained capacity can emphasise their 'exclusivity' as the bulk of regional expansion takes place on the mainland. The mainland connection is also a potential strength as the islands attempt to develop closer economic and theme-based linkages with their hinterland.

The Whitsunday islands have previously focused their promotional campaigns on the proximity of the Great Barrier Reef. Developments in water craft technology have brought the Reef closer to the islands and World Heritage listing for the Reef has helped reinforce the Whitsundays' claim to be 'unique'. The activities of the Marine Park Authority have also provided professional environmental management to underpin the natural resource credentials of the region – it is too easy for promoters to emphasise the apparent natural beauty of an area without taking account of medium- to long-term environmental problems.

The Whitsundays do exemplify one problem common to 'exclusive' islands. They maintain relatively little contact with mainland residents and virtually no contact with the indigenous population that once inhabited the region. The environmental context provided by the resorts has focused predominantly on the physical realm. Closer linkages with the mainland offer the prospect of stronger social and political integration. Aborigines could be employed as park rangers and guides, in both the interpretation of sites of Aboriginal significance and in overall environmental management. Local arts and crafts from the Whitsunday mainland could be given greater prominence. Building this social and cultural dimension will strengthen the region. Other island resort destinations should take note of the prediction that ignoring social, cultural and community linkages may lead to the undermining of the appeal of the destination.

The Whitsunday resorts are relatively large and a number are managed by transnational hotel corporations. This has provided the region with access to management and marketing expertise, but has contributed to the crowding out of smaller informal resorts. Though the larger corporations are skilled at managing 'relaxation', their resorts exhibit an element of institutionalisation not evident in the Mamanucas. The high – some might say overly elaborate – quality of facilities developed at resorts such as Hayman, Club Med, Hamilton and Daydream (at a huge loss to the initial developers and owners) was symptomatic of the excesses of the 1980s, but now provides the region with a highly competitive infrastructure for both the domestic and the international markets. Resort developers in other parts of the world should, however, take careful note of the huge losses made by developers in the Whitsundays. Island destinations located far from key metropolitan markets and heavily dependent on scheduled air transport are highly risky ventures and investments in such settings should be approached with caution. The apparent marketing successes of the Whitsundays should not obscure the boom and bust nature of property development in such settings.

For the Whitsundays region, the initial resort over-capitalisation becomes a problem only if the current operators and owners are unable to maintain the facilities at an appropriate level. This is unlikely to happen as long as international hotel management companies compete for the opportunity to manage properties in the islands. Fortunately for the region, it was predominantly interstate investors who funded the extravagant expansion of Whitsunday resort tourism, saving the locals from misfortune during periods of loss-making! It does not, however, assist the creation of a stable investment climate.

What can the lessons of the Whitsundays show island resort destinations in other parts of the world? Firstly that the projection of a coherent regional image is likely to provide operators with benefits and with direction. The promotion of individual resorts – notably Hayman and Hamilton islands – as separate from and superior to the rest of the region, proved counterproductive. Individual properties do undoubtedly attract distinct markets and should maintain their different positioning in the marketplace. However, the sacrifice of *context* in the pursuit of superiority is ill-advised as the resort managers appear to have acknowledged. Consumers are usually well placed to form a hierarchy of relative resort appeal for themselves.

Developers, managers and marketeers should not allow themselves to be seduced into the idea of the primacy of the deserted, stand-alone tropical island. The loss of market share experienced by South Pacific islands in the Australian market may be symptomatic of the failure of the various island destinations to communicate an appealing context of diverse cultures and environments.

The Whitsundays demonstrate the potential benefits of combining the attractions of a group of islands, with those of an adjoining mainland. This may not *always* apply to all equivalent destinations, but the Whitsundays have shown that the exclusivity of islands can be juxtaposed with the accessibility and budget consciousness of the mainland properties. To date, the image of overall appeal (including exclusivity) associated with the Whitsundays appears to have been maintained in the Australian market. This success of projecting the availability of a range of properties, catering to a wide range of tastes and a wide range of budgets within an overall image of desirability, is worthy of note for other island resort destinations though, as previously stated, the provision of more luxurious resorts was brought about at the expense of initial developers.

Like the Whitsundays, the Mamanucas region is fairly well defined but in contrast it has a weak institutional base for marketing and development with the poorly resourced local Hotels Association branch shouldering all of the burden. The FVB allocates co-operative marketing funds but not to tourism regions in particular. This contrasts with current practice in Australia, where substantial State and Commonwealth funds have been used to encourage tourism regions and where strong and viable tourism regions are regarded as a vital underpinning to a viable tourism sector. Given the lack of funds and leadership from the national authorities, the various Mamanuca initiatives have been sporadic and supported by only a core of operators. The relatively small scale of the resorts makes the need for collective action more urgent. Island destinations such as Fiji urgently need to acknowledge the growing importance of regional tourism. The fact that developing countries are short of resources is no excuse. A strong regional base is vital for a strong and internationally competitive industry.

As is the case in the Whitsundays, the Mamanucas are a relatively exclusive haven for higher-spending visitors compared with the mainland. However, there is a lack of a strong institutional linkage with the other 'exclusive' areas which adjoin the group, namely the Yasawa Group to the North and the Denarau Resort properties to the South. Participation by these resorts in initiatives involving the Mamanucas would create economies of scale and could help raise the profile of the group.

Such co-operation will require the Yasawa and Denarau properties and corporate interests to take a stronger involvement, possibly at the expense of the pioneering entrepreneurs who first developed the area. This would mirror what has occurred previously in the Whitsundays, according to

Barr's historical account of that area (1990). ITT Sheraton might take an interest since it has properties on both Denarau and in the Mamanucas/Yasawas (Vomo). The process could be encouraged by the Fiji Government as a model for regional tourism, though clearly the willingness of key players in the region to participate would need to be tested. The Whitsundays example has shown what can be achieved by the potent combination of three key factors: involvement by international chains, a blending of island and mainland properties and strong government support. French Polynesia's successful 'Tahiti and her Islands' campaign has shown how a developing Pacific country can facilitate the emergence of strong subregional island 'brands'. The issue of branding at regional level is worthy of further investigation, preferably involving a comparison of various different countries.

The Mamanucas lag behind the Whitsundays in environmental management. Whilst the relatively small scale of development and existence of some voluntary schemes have meant that water quality and other key environmental measures are by no means at crisis point, the absence of any formalised network of national parks and protected areas in Fiji constrains the emergence of strong environmental tourism concepts. The warning of the 1989 Tourism Masterplan (Coopers and Lybrand 1989) about tourist impacts and densities in the Mamanucas points to the urgency of addressing environmental challenges. Informal arrangements such as those practised with the local landowners by Beachcomber and Treasure Islands require urgent formalisation. An officially-designated protected area would provide the Mamanucas with an environmental attraction capable of drawing visitors to the region. Though on a larger scale, the promotion of the Great Barrier Reef by the Whitsundays has shown the potential of such natural attractions, even for the apparently hedonistic sun, sea and sand market. In other parts of the world, not all island resort regions have natural attraction drawcards on the scale of the Great Barrier Reef. Such destinations have more in common with the Mamanucas, torn between the option of developing a major tourism attraction or expanding the range of smaller-scale activities and attractions made available for guests.

It may be that because of differences in scale, the two regions need to adopt different models for the development of tourism in tandem with sustainable environment concepts. As a major natural icon, the Great Barrier Reef and its location within a marine park helps perpetuate large-scale, resort-based tourism, since the self-regulatory approach lends itself to participation by large-scale, purportedly responsible corporate citizens (Craik 1987a). In the Mamanucas, the smaller scale of resorts may necessitate a more co-operative model between the various resorts with prompting from the new Fiji Department of the Environment. The region could certainly lend itself as a pilot project for the new Department involving resort sponsorship of protected areas in a form of public–private partnership. Such partnerships between government departments, individual resorts and tourism regions could be a viable model for other developing countries.

Fiji is ahead of the Whitsundays in its integration of the host community. The economy of the Mamanucas includes fishing, agriculture and arts and crafts as well as tourism. The relationship with landowners is formalised through the NLTB lease payments, through the employment of staff from the landowning unit and through close liaison and co-operation with the local chiefs. The restriction of tourism development to low-rise bures has been vindicated by its success in minimising the otherwise obvious contrast between guests and local residents. Freestanding bures near the beach convey an informality with obvious appeal to holidaymakers pursuing their ideal 'Pacific paradise' and appear a suitable style for smaller-scale resorts. Whether the approach can work on a grander scale will become evident as Mana Island proceeds with its major expansion and possible tripling of bure capacity, although the risk of losing the intimate atmosphere currently experienced should be acknowledged. Without the type of development proposed for Mana, however, the Mamanucas risk perpetuating their current reputation as a destination suited largely to families with young children and lacking in appeal for teenagers. For the Whitsundays, the unfavourable comparison with the Mamanucas over community integration points to an issue that will grow in prominence across Australia as the impact of the Mabo decision continues to be felt and support for 'community-based' tourism grows.

To what extent can the Mamanucas act as a model, albeit an imperfect one, for the development of island resorts elsewhere? The resorts have successfully exploited their proximity to an international airport, through the provision of diverse transfer alternatives (launch, helicopter, sea-plane and scheduled aircraft). They have been less successful (as have the national marketing authorities) at *communicating* their accessibility.

For other developing countries, the Mamanucas demonstrate a practical application of a tourism masterplan (UNDP 1973) that took a strong stand concerning the scale and aesthetics of resort development. Whilst some critics may rightly challenge the assumed 'authenticity' of thatched bures on the beach, that characterise the Mamanucas, the concept has continuing merit in the 1990s. The same cannot be said for the high-rise concept behind Hamilton Island. Despite its coherence in the Mamanucas, the 'small is beautiful' philosophy has major challenges associated with it. The perception amongst travel agents and tour operators that there is 'nothing to do' in the Mamanucas needs to be addressed by operators.

In examining the impact of tourism industry structure on the two regions, it has been concluded that they have benefited less from airline deregulation than other destinations such as Far North Queensland because of their dependence on leisure travel revenues. The Whitsundays have at least experienced an expansion of capacity and have shared the spoils of an overall reduction in domestic air ticket prices. Though the monopoly of access through Hamilton Island Airport by one airline, Ansett Australia, has been a constraint, the expansion of Whitsunday Airport may bring about a more competitive environment. The Mamanucas have also been constrained by the lack of airline competition between Australia and Nadi, with Qantas and Air Pacific's code-sharing agreement symptomatic of the close association between the two carriers. The decline in competition was hastened by the drastic reduction in the number of stops by trans-Pacific flights in Fiji.

Unfortunately for the Mamanucas, the demand for main annual holidays on the beach in the South Pacific has fallen in Australia and Fiji has yet to position itself as a short-break destination. The authorities have not yet acknowledged the significant competitive advantage that the country possesses in this market. Equivalent island resort destinations in other parts of the world could learn from Fiji's slow response to this vital issue.

Ansett dominates the Whitsundays more than Air Pacific does the Mamanucas because of the active holiday division of the Australian-based group. Most passengers bound for the Whitsunday islands through Hamilton Island Airport have purchased an Ansett holiday. Air Pacific does not possess a tour operating programme, though the promotion of Air Pacific packages in mid-1996 may foreshadow the emergence of a wholesaling programme. Until this occurs, the Fiji resorts will be dependent on a plethora of tour operators, some of them quite small and a number scarcely profitable. Competing with the many small operators, Qantas is by far the largest holiday operator with its Jetabout and Viva! brands, recently amalgamated into Qantas Holidays. Because of their dependence on tour operator intermediaries, the Mamanuca resorts have insufficient contact with travel agents, especially those who are members of multi-outlet chains. As Poon (1993) has observed, resorts throughout the world will need to take greater control over their own marketing either directly or through travel agents, rather than relying exclusively on tour operators to do their marketing for them. This dependence on tour operators plus an overall dearth of marketing resources has led to a shortage of information provided to consumers apart from basic brochures highlighting the actual resorts. Consumers perceive that the Mamanucas are remote, inaccessible and have little of cultural interest. Unless more funds are allocated to marketing Fiji as a whole, the existing tour operators with their small margins and declining markets may be reluctant to engage in the type of market segmentation necessary to attract special-interest tourists, never mind ecotourists. They will tend to concentrate on their own survival, rather than facing the longer-term challenge which Fiji must address.

The increasing concentration within the industry is a dilemma for the Mamanucas. Although Qantas has recently announced a proposed sell-off of its island resort division, the company is still more preoccupied by its Queensland properties than by its Fiji connection. The recent reduction in its flights from Australia to North America via Nadi and Hawaii is sympto-matic. Fiji and the Mamanucas may need to specialise as a destination or risk becoming more dependent still on a lukewarm Qantas and its sub-sidiary holiday brands. One option might be for a strategic alliance between Air Pacific and specified tour operators, preferably one that is Fiji-based and hence dedicated to the destination. Another option would be to bring about greater competition between Australia and Fiji through the com-mencement of services by another carrier such as Air New Zealand or Ansett. Otherwise, the Mamanucas risk being further marginalised as competition from Australian and Asian resorts intensifies.

Structural differences between international and domestic travel have lessened. An example is Qantas' implementation of its resort classification and grading scheme. This scheme was previously confined to its overseas destinations (by Qantas Jetabout Holidays), but now applies to the products of the domestic Qantas Australian Holidays brand. Most of the changes introduced by Qantas have benefited domestic resort destinations such as the Whitsundays, though not directly in the case of the Whitsundays. Since Brampton is the only resort in the vicinity of the Whitsundays that is Qantas operated, the extra promotional dollars spent by Qantas on its Queensland properties have probably transferred business to other parts of the Great Barrier Reef. On the other hand, Ansett has responded to the various Qantas initiatives by boosting its Queensland Island promotions. Since most of its island resort interests are concentrated in the Whitsun-days, the result may have been to benefit the region indirectly. It is clear that at least some of the growth of interstate travel to the Whitsundays has been at the expense of Fiji. The move by Qantas to streamline administration and to reduce duplication by combining key aspects of its domestic and international programmes should be instructive for countries which have previously operated separate domestic and international structures. Cost pressures and increasingly deregulated international competition are likely to hasten the occurrence of such rationalisation. Vulnerable island resort destinations should take close note of such potential changes.

The different style of industry organisation prevailing in the two regions is instructive. The success of the Whitsundays in tapping the short-break market is something that the Mamanucas might well emulate. Within the region, Air New Zealand now promotes its own short-breaks programme aggressively in the Australian market (to major cities and tourism centres). The recent relaxation of the rules governing airfares between Australia and Fiji should allow tour operators greater flexibility to introduce tactical marketing initiatives and travel agents to sell short breaks in Fiji at relatively short notice. International short breaks have been popular in Europe for many years and suggest that a move by Fiji could be successful, despite hesitancy on the part of the FVB and certain Mamanuca managers.

The Whitsundays and Mamanucas exemplify how susceptible island resorts are to structural changes in the air transport sector, where they are dependent on scheduled flights. The Whitsundays experienced significant losses during the Australian air pilots' dispute with the island properties worst affected. In the case of Fiji, the preoccupation of Qantas with geopolitical concerns reduced the destination to marginal status. Were it not for the reasonable strength and capacity provided by Fiji's international carrier Air Pacific, the impact would have been greater still. The experience of both areas suggests the need for contingency planning by destination regions to take into account potential changes to the direction of air transport.

This study has attempted to provide a bridge between the imagery of resort promotion and the social, environmental and industry dimensions of resorts. Neither Fiji nor the Whitsundays conform to the classic South Pacific myth, based as it is on concepts of Polynesian rather than Melanesian or Aboriginal culture and settings. Nevertheless, the study identified a prolific literature dating from the nineteenth century which highlighted the tropical delights of both the Queensland and Fiji islands. Like those in the South Pacific, the Queensland islands have their own mythology, founded on literary and related associations. Whereas Queensland islands have borrowed from the South Pacific myth (for example, the widespread use of the label 'Polynesian bures'), the Pacific islands do not appear to have borrowed from the Queensland myth. Future promotions by Fiji might benefit from acknowledging the competing and increasingly effective messages being projected by their Queensland equivalents.

Resort imagery is placing an increasing emphasis on the subjective with promotions typically urging consumers to picture themselves as the centre of the world. This world is characterised as a land of indulgence, far away from the problems (or realities) of everyday life. Such images point to an increasing 'sophistication' in advertising and to its increasing persuasiveness. Whether sophistication means 'better informed' as it is typically portrayed in the marketing literature, or 'manipulated', is debatable. Resort marketing is an example of the increasing emphasis on the commodification of *experiences* rather than on tangible consumer goods. This style of resort promotion is symptomatic of increased 'lifestyle-based' persuasion and associated consumption. As a follow-up to the present study, a detailed evaluation of promotional messages using content analysis could be a useful means of arriving at some quantification. The extent to which words such as 'paradise' are employed and the use of metaphors to convey the prospect of 'fun', would also be worthy of further study.

The clienteles of the two island groups are becoming less Australian and more international. Fiji's share of Australian outbound travel has dropped and the proportion of tourist arrivals accounted for by other nationalities has grown. Focus group participants expressed a preference for resorts with a diverse international clientele and it is clear that properties such as Daydream, which have been predominantly Australian, need to acknowledge these preferences. Tourism in the South-West Pacific is increasingly cosmopolitan. While this trend should be welcomed by the resorts, the need to respond to the preferences of different source markets should be accommodated. The Whitsundays, for example, should not over-emphasise the provision of facilities for young children to overseas visitors, since fewer international visitors travel with children. In the case of the Mamanucas, the less price sensitive, long-haul markets which have been embraced wholeheartedly do not offer the same repeat visitation prospects and should not be allowed to expand at the expense of the Australian market.

Growing cosmopolitanism is typical of resorts in the Asia–Pacific region. Whereas the dominance of the European markets for Mediterranean islands and the US market for Caribbean islands goes relatively unchallenged because of the large populations of the key generating countries, this cannot be said of those countries which have traditionally been a part of Australia's 'pleasure periphery'. Australia's rate of economic growth has lagged behind that of its Asian trading partners over the past fifteen to twenty years, resulting in a reduced market share for Australians. Simultaneously, American and European tourists are travelling long-haul in increasing numbers. The result has been a changing and less significant role for Australia as a tourist-generating country, despite the bullish forecasts of the Tourism Forecasting Council (TFC 1996). The Whitsundays and the Mamanucas exemplify the stronger Asia–Pacific focus of tourism resorts within the region. They have highlighted some of the strains experienced by vulnerable island resorts in the face of a shifting balance between the dominance of short-, medium- and long-haul markets.

Both island groups have emphasised the family market. Always a mainstay of the Mamanuca resorts, families have been embraced by some Whitsunday resorts more recently, with these properties moving quickly to install 'state of the art' childcare facilities. Mamanuca resort managers should monitor these developments closely. While over-dependence on families may not be good business practice, the core status of this market should be acknowledged. It is easier and more economical to retain existing consumers than to attract new ones (Baker 1985). Resort managers in other parts of the world should take note of the Whitsundays' apparent success in adopting 'state of the art' childcare facilities as a part of their product mix.

Both island groups are exclusive operations relative to their respective regions. Social segregation occurs as lower-budget travellers opt for the cheaper 'hub' locations and never reach the more exclusive island periphery. Further segregation takes place when consumers select an individual resort, though discounting by even the most exclusive (Hayman) has softened this impression. Once on the island, the social atmosphere is ostensibly unstratified with social distinctions kept to a minimum. As evidenced by Hayman's abandonment of formality, an air of relaxation and apparent egalitarianism is an important ingredient for resorts, irrespective of their target audience. It is difficult to judge whether Hayman's retreat from formality is an acknowledgement of the special expectations of the Australian market (informality). Prospective developers, in Australia at least, should take note of the Hayman experience.

The present study has not been able to prove that culture- and environment-related activities will be demanded by holidaymakers who in previous years would have been content with relaxation, some socialising and watersports. There are, however, clear opportunities to enhance the holiday experience in certain resorts by showcasing the culture and environment of the destination in a more systematic way (through, for example, improved availability of regionally produced retail products). Island resort guests may not crave the commercial delights of shopping malls but they share the acquisitive urge of most holidaymakers to secure *souvenirs* (including items of clothing) reminiscent of the destination. Neither island group currently caters adequately for such demands. Achieving a balance between authenticity and commercialism will be a key challenge for island resorts everywhere.

The study has found that resort developers and managers would be well advised to consider the resort as a complete community reflecting a variety of individual (i.e. tourist) aspirations. The concept should also incorporate the aspirations of staff who either participate in the community themselves as residents (with accommodation provided by the resort), or share their culture and lifestyle with the tourists (in the case of those who live in nearby settlements). In the case of Club Med, the staff input is deliberately international, whereas in Fiji the emphasis is more narrowly focused, with one group of the national population (Fijians) considered more suitable for customer contact roles than their Indian fellow-citizens. The concept of island resorts as complete communities merits further research and is worthy of greater emphasis in hotel management texts which set out to be relevant as a guide to resort operations.

Resort managers will need to behave less like hotel managers confronted with additional facility management responsibilities and become more like community mayors taking an interest in the 'welfare' of their constituency. They will need to decide how much artifice is appropriate for their clients – do they offer pure fantasy as the Hawaiian resorts of Christopher Hemmeter attempt to do, or do they draw upon local culture to differentiate themselves? What if there is no surviving local culture as in the case of the Whitsunday Islands? Are they to become a 'home away from home', or something as far removed as possible from mundane everyday life? Examining the role of resorts as 'communities' has provided insights into all of the preceding questions. Not all resorts need to pursue an ideal of 'authenticity' based on some interpretation of 'tradition'. While fantasy themes may be an option for Whitsunday resorts such as Hamilton Island, with its highly commercialised ethos, it is suggested that the rich cultural heritage of the Pacific lends itself to resorts which pursue authenticity wholeheartedly. The absence of current Aboriginal settlement in the Whitsundays deprives the region of any depiction of authenticity based on a contemporary human involvement. Nevertheless, an acknowledgement of Aboriginal involvement in the historical shaping of the region and recognition of contemporary Aboriginal issues in Queensland could add an additional dimension. It is ironic that Aboriginal and Torres Strait islander entertainers were in fact active in the immediate post-war years. This phenomenon itself may be worthy of investigation for possible highlighting as an element in the historical development of Whitsundays tourism.

'Environmental context' is the key feature which differentiates resorts from other types of consumption places such as theme parks and shopping malls, and resort operators must be acutely aware of issues concerning both micro- and macro-environments. Whether this should involve the full-blown 'ecotourism' option is debatable, since ecotourists constitute a small specialist consumer group, many of whom would regard larger resorts as anathema. Despite the criticism levelled by some sociologists that typical depictions of the natural environment have commodified the landscape and led to the 'packaging' of public space within the boundaries of the camera lens, the present research has proposed that resorts themselves should be acknowledged as a 'real' feature of the landscape. Resorts can either look inwards, becoming 'enclave' resorts, or can open up to the adjoining environment. Clearly articulating the linkages and interdependencies with the natural environment is central to the latter approach. This applies in both the Whitsundays and the Mamanucas, despite the very different levels of community integration (see figure 9.1).

Enclave tourism in its pure form may be a theoretical option but almost all resorts will in practice look outwards, even if only to exploit the resource for the 'benefit' of their guests. This approach appears to have merit, even for operators committed to maximising the proportion of the tourist dollars spent on services offered by the resort. It is not in the interests of resort managers that tourists feel 'cooped up' and they can earn useful revenue through organising trips into the adjoining environment themselves. The bay may be the setting for jet-skiing, the air for para-sailing and the ocean for discharging effluent. In contrasting inward- versus outward-looking resorts, we should consider how the resort owners and operators interpret the value of the surrounding environment. Resorts will ultimately make the best contribution to the locality, and to the region in which they are located, by actively investing in local institutions and in socio-cultural and environmental concepts such as marine reserves and village tours. This will provide

Figure 9.1 'Neighbourly' Club Med?
Club Med's global network has been the epitome of 'enclave' tourism. Is Lindeman making a half-hearted claim to integration (note the absence of people)?
Source: Nigel Stoker, Managing Director, Club Med (courtesy of Club Med)

local populations with a sense (real or manufactured) of ownership and will enable the resort to move beyond a pure consumer fantasy product to one which can offer visitors a sense of context. According to the typology developed by Budowski (1976), resorts could potentially take one of three options for their relationship with the environment – conflict, co-existence or symbiosis. The role of a stable and locally-based workforce may play an important role in bringing about the last of the three alternatives. The latter is clearly preferable in the Whitsundays and the Mamanucas.

Though certain medium- to long-term environmental concerns are evident, the immediate challenge for the economic viability (and hence sustainability) of Fiji's tourism industry is to address its dramatic drop in market share of the Australian outbound travel market. As stated previously, this fall is attributable to a mixture of controllable and uncontrollable factors. International competition will continue to intensify. Fiji can, however, improve its positioning in the Australian market by reinforcing its 'Fiji Islands' concept (which recognises that Fiji is more than Viti Levu and consists of other island groups such as the Mamanucas) and by focusing on particular market segments, as was done successfully in the 'Dive Mamanucas' initiative. Such specific targeting of special-interest groups needs to be more frequent and more systematic. Targeting other special-interest groups may entail physical modifications to some tourist facilities (for example, conference and meeting rooms) as well as greater incorporation of natural and cultural heritage into the tourism experience that is offered (FVB 1993a).

The problems being faced by Fiji are typical of those encountered by many resort islands located at the margins and lacking adequate resources for promotion and product development. Though the present research has emphasised the need to develop local/regional concepts in areas such as the Mamanucas, there may be some value in considering a total South Pacific promotional initiative in Australia. Though this would not negate the need for product development, it would at least mount a coherent challenge to the growing predominance of Asian destinations for beach holidays.

In the Mamanucas, the existing visitor experience is more than a simple blend of 'sun, sea and sand' and those who reduce it to these three elements are underrating its complexity. The unobtrusive nature of the resorts, the channelling of some resort revenues to the community, the performance of *mekes* and other cultural spectacles by local residents, the small scale of most resort developments and the highly personalised service by Fijians from the immediate vicinity have all helped to absorb tourism into the life of the community. The larger scale of the Whitsunday properties, their management by international chains and the absence of an indigenous population has made tourism less culturally and socially integrated than in the Mamanucas. Nevertheless, in any attempt by the two groups to bolster the environmental component of the experience, the Whitsundays start with a key advantage because the relevant resort islands are designated National Park and the nearby Great Barrier Reef is managed under the strict requirements that accompany World Heritage listing.

Given that this relatively high level of integration already exists in the Mamanucas, what is currently preventing the expansion of the natural and cultural heritage elements still further? One is resort manager scepticism. A number questioned whether Australian consumers were genuinely interested in ecotourism and expressed the view that the Fiji Department of Tourism had pursued the initiative over-zealously, overemphasising activities in more remote parts of the country. Other managers attributed the 'Australian problem' to marketing and not to any need to alter the existing core visitor experience which they saw as relaxation. Naitasi was the exception in attempting to diversify and reposition its product amongst its existing clientele as well as attracting more special-interest groups with a particular interest in the environment.

The Mamanucas need to observe caution in proclaiming their environmental credentials, since the area might fail to deliver on any promises that it makes of a pristine environment. The Mamanucas are small, fragile islands lacking the environmental controls and monitoring procedures necessary to cope with larger visitor numbers. This is a risky competitive situation vis-à-vis the Whitsundays. Given the potential problems arising from current environmental management practice, it is advisable that cultural heritage attractions in the Mamanucas receive equal priority with the issue of 'pristine environment'. Perhaps the human dimension of the Fiji smile, which has formed the cornerstone of previous FVB promotions, could be supplemented with the friendly Fijian or Indian keen to share his or her cultural heritage with the visitor. In view of the close connections between the resorts and the island communities, this sort of message is a practical one which can be delivered. It is of course an important prerequisite that cultural assets must first be thoroughly identified and interpreted and that tour guide training of the type currently under way at Naitasi takes place.

The Mamanucas must diversify their existing Australian market by adding an extra experiential dimension for those currently attracted by 'sun, sea and sand'. These markets can be supplemented, particularly in off-peak periods, by those with a more explicit interest in natural or cultural heritage such as the Earthwatch scientific organisation. Such highly targeted initiatives, involving small numbers of special-interest visitors, are unlikely to be adopted, at least during their early development, by mainstream tour operators preoccupied with volume. A collective approach by the Mamanucas could work, relying on the type of direct marketing methods used in the 'Dive Mamanucas' campaign and using a collaborative approach. Leadership is needed at the national level to provide regions with incentives to develop such special-interest markets.

The effective target marketing done by the Whitsundays and other Queensland islands to the dive and honeymoon markets (S. Brewster pers.comm.) shows the benefits of pooling resources and careful market segmentation. Some examples of product adoption/modification to meet the needs of particular target markets are already taking place at individual

Mamanuca resorts. More could be achieved still, if the Mamanucas were to work together in their segmentation, either direct to the consumer or via tour operators. One Fiji-based operator, Rosie's Tours, has shown a particular interest in special-interest marketing and could be used for concept development. Any strategy, however, must avoid a preoccupation with 'new product' and with 'new markets'. The basic perception of the Mamanucas as a somewhat remote area lacking diversity, and as not offering a 'real cultural experience' compared to popular destinations such as Bali, needs to be addressed first. In other words, such island resort destinations need to emphasise their existing, core markets – and the service that these groups receive. Diversification of products and markets is costly and best approached gradually.

In Fiji, debate over the future of tourism risks becoming polarised between the supposed alternatives of ecotourism *or* sun, sea and sand tourism and between the 'old' tourism and the 'new' tourism. The Mamanucas have always been more than the three Ss. The group should embrace the prospect of greater incorporation of cultural and natural heritage elements in their tourism, whilst not abandoning their long-established family market strategies. Ecotourism may be tempting for the Mamanucas as an apparent bridge between the high values attached to adjoining environments and the need to rectify recent falls in Australian arrivals, but the scale of ecotourism and special-interest markets will not address the fundamental problem.

This study has shown that as an example of the various types of post-modern 'consumption places' cited by Sack (1992), island resorts are the most highly dependent on conditions in the adjoining environment and on perceptions of the state of such environments. This is a dilemma for tourism regions such as the Mamanucas and the Whitsundays because both need pristine environments for their continued attractiveness and viability as tourism destinations. Healy's term 'common pool' land is a good description for land in the immediate vicinity of resorts but outside the area covered by the resort lease.

Reluctance to participate by private operators occurs less in the Whitsundays because the bulk of the adjoining environment is managed by the GBRMPA. Individual operators are willing to invest in facilities such as coral viewing platforms on public land or water, in partnership with the public sector because they do not regard such investments as benefiting their competitors directly. Such initiatives provide a stable institutional framework for the management and regulation of both marine and terrestrial environments. The equivalent public authorities in developing countries such as Fiji suffer from a lack of political will and a shortage of resources. This places the Fiji resorts at a competitive disadvantage, especially as there is little incentive for private operators to become involved in environmental management issues. It is clear that the Mamanucas cannot replicate the conditions which prevail in the Whitsundays. A body equivalent to the GBRMPA is unlikely to be established in Fiji, at least in the short term. The present research has, however, shown how important the overall environmental context is becoming as a determinant of competitive advantage. Mamanuca and other Fiji resorts must conceive of ways in which the institutional disadvantage can be minimised, probably in consultation with international funding agencies. The development of a visitor attraction which complements the adjoining environment such as an aquarium might be an appropriate means of showcasing the destination and adding to the range of visitor activities. Ayala's suggestion (1991b) that resorts 'adopt' an adjoining natural attraction and then contribute to its maintenance is a sound one, though one which individual operators will be unlikely to undertake without co-operation amongst competitors and without the provision of outside funding. Tourism destinations in developing countries should observe the advantages possessed by developed countries such as Australia with their framework of environmental management and ensure the strategic security of their own environmental attractions.

It is probably counter-productive to urge a blueprint for island resorts. In the two regions of the Whitsundays and the Mamanucas, diversity within a coherent framework relevant to local needs is the most likely to yield success. A key is to recognise the limits to growth. In the Whitsundays this means a focus on mainland development, rather than on island development. Investment in the Whitsundays has been less of a problem than fragmented marketing. In the Mamanucas, lack of investment has been a problem and critical decisions are needed about how much development is appropriate. Such assessments are likely to be most effective when pursued within an accepted regional context (with clearly accepted and demarcated boundaries).

Queensland in general and the Whitsundays in particular have shown a stronger commitment to such a process than Fiji, with strategic tourism plans in place at both State and regional (Whitsundays) level, though both have been delayed in progressing from draft to final form which is a cause for concern. The absence of any tourism plan in Fiji, relevant to particular regions such as the Mamanucas, is unfortunate. In the absence of either a plan or an institutional structure at regional level, the Mamanucas may be condemned to a spiral of limited resources and fragmented activity. This dilemma is not atypical in developing countries, but is particularly unfortunate for Fiji in view of the many positive results which arose from the 1973 Plan. Other developing countries in particular would be well advised to ensure that the tourism planning process gives due attention to established island resort regions. This is notably the case for destinations which are becoming marginalised by political and aviation trends.

Since Boorstin (1964), many analysts have been disparaging of the meaninglessness and vacuousness of tourism, compared to the more adventurous activities of individual travellers. Resorts have been depicted as the epitome of the meaningless experience (Krippendorf 1987). The present research has challenged such assumptions. In attempting to assess the best prospects for destination regions, it has been shown that island resorts can play a constructive and central role. They help to manage the impact of concentrated visitor numbers and offer the prospect to showcase the destination region, including social, cultural and environmental dimensions. The view that island resorts represent no more than a stereotyped blend of sun, sea and sand is outdated. Sensitively managed resorts can complement the landscape. They also offer their guests the prospect of a shared community experience alongside elements of secluded tranquillity. Island resorts which move beyond hackneyed claims to be the ultimate tropical paradise can be at the forefront of the *new tourism* and not simply outcasts of the now *old tourism*.

By concentrating its attention at local/regional as opposed to national level, this study has provided an alternative perspective to the growing number of studies focusing on island nation destinations generally. Further research could examine a wider range of small island destinations. Such research could examine the destinations from the perspective of a single generating country, or could use a range of criteria to evaluate the chosen destinations with reference to a variety of different source markets. This research could usefully involve a multidisciplinary team, to ensure a thorough examination of environmental, socio-cultural and business issues and to provide an extra range of perspectives not available to a single researcher.

Another direction for future research could involve a range of coverage given to small island destinations in the process of preparing and implementing masterplans. Such plans have been drawn up for a variety of island nations including many in the South Pacific, but cross-country comparisons are rarely undertaken. In undertaking such comparisons, the focus could either be on a particular dimension (for example, environmental management for resorts), or using a holistic approach such as has been attempted in the present study.

This study has begun to explore the idea of whether a domestic island holiday is a different concept from an international island resort holiday from the consumer's point of view. It has been shown that the product offered is relatively comparable and substitutable and hence that domestic and international destinations do indeed compete directly for a very large proportion of travel consumers. With the emergence of island resorts in South-East Asia and the growth of tourism markets in these countries, it could be interesting to study the competitiveness of resorts from the point of view of emerging markets such as Thailand, Malaysia and Indonesia (King and McVey 1996b). These countries are becoming relevant as potential consumers on the one hand, and as potential competitors on the other, for island resorts in all parts of the world.

MacCannell's pioneering work *The Tourist* (1976) used the field of tourism to exemplify the key social issues facing contemporary humanity. Using a more empirical approach, this study has shown that island resorts are consumption places which exemplify key features of the postmodern condition. For resort managers this means that island resort regions must capitalise on their sociocultural and environmental settings to satisfy the changing holiday expectations of travellers. As outposts of the global village, island resorts will need to be alert to changing societal and consumer expectations if they are to remain both attractive and financially viable.

APPENDIX
List of individuals consulted

THE WHITSUNDAYS

(i) Structured personal interviews (where taped, marked **)

Brewster, S. – Product Development Manager, Sunlover Holidays, 9 February 1994 **

Cogar, S. – General Manager, Hamilton Island Resort, 10 November 1993 **

Collins, K. – General Manager, South Molle Resort, 6 November 1993 **

Giampaolo, B. – Chef de Village, Club Med Lindeman Island, 4 November 1993 **

Gregg, S. – General Manager Marketing, Queensland Tourist and Travel Corporation, 9 February 1994**

Harrold, N. – Marketing Manager, Group Sales Australia, Qantas Airways, March 1992

Hutchen, D. – Chairman, Whitsunday Visitors and Promotions Bureau and Chief Executive, Fantasea Cruises, 1 November 1993 **

Kelly, N. – General Manager, Radisson Long Island Resort, 3 November 1993 **

Klein, T. – General Manager, Hayman Island Resort, 5 November 1993 **

Lindsay, R. – Marketing Manager, Page McGeary Resorts (including Palm Bay Hideaway)

Mahony, G. – General Manager, Brampton Island Resort, 8 November 1993 **

Maloney, C. – Joint Resident Manager, Palm Bay Hideaway, 2 November 1993 **

Maloney, J. – Joint Resident Manager, Palm Bay Hideaway, 2 November 1993 **

Robertson, G. – Manager, Ansett Holidays, March 1992

Tall, C. – Resort Manager, Hook Island Wilderness Lodge, 6 November 1993 **

Vincent, W.L. – General Manager, Daydream Island Travelodge Resort, 9 November 1993 **

Wallace, B. – Regional Tourism Co-ordinator, Queensland Tourist and Travel Corporation, 9 February 1994 **

(ii) Informal discussions/interviews about the research were conducted with the following individuals

Bennetts, R. – General Manager, Coral Sea Resort, 5 November 1993

Clarke, D. – Chief Executive Officer, Jetset, October 1991

Court, G. – Manager, Market Research, Planning and Administration, Ansett Australia, September 1994

Davie, R. – Office of the Co-ordinator General, Government of Queensland, 20 November 1994

Diamond, B. – Guest Relations Manager, Hamilton Island Resort, 10 November 1993

Herridge, N. – Product Co-ordinator, Ansett Holidays, October 1990

Kelly, N. – Guest Relations Manager, Hayman Island Resort, 5 November 1993

Lilley, J. – Distribution Manager, Australian Airlines, October 1990

McKinnon, M. – Director, Hambleton Ruff Advertising, September 1993

Millmore, T. – Chairman, Australian Council of Tour Wholesalers, March 1991

Stoker, N. – General Manager, Club Med Australia, October 1993

Wall, S. – Tourism Development Co-ordinator, Tourism Policy Unit, Queensland

THE MAMANUCAS

(i) Structured taped personal interviews

Bakaniceva, N. – Assistant Manager, Senior Estate Officer, Native Land Trust Board, 2 February 1994

Cabaniuk, S. – Planning Adviser, Native Land Trust Board, 2 February 1994

Dreunimisimisi, A.V. – General Manager, Resort and Cruises, Beachcomber Cruises Ltd, 31 January 1994

Dutta, E. – Director of Marketing, Air Pacific, 1 February 1994

Erbsleben, P. – Managing Director, United Touring Fiji, 31 January 1994

Francis, J. – Principal Tourism Officer (Ecotourism), Department of Tourism, Fiji Government, 1 February 1994

Gucake, M. – Director, Department of Tourism, Government of Fiji, 1 February 1994

Lolohea, S. – Head of Marketing and Promotion, Tourism Council of the South Pacific, 3 February 1994

MacDonald, M. – Manager, Naitasi Resort, 27 January 1994

MacDonald, S. – Co-Manager, Naitasi Resort, 27 January 1994

Palise, K. – Resort Manager, Mana Island Resort (Fiji) Ltd, 29 January 1994

Palmer, N. – General Manager, Sheraton Vomo Island Resort, 30 January 1994

Reed, A. – Managing Director, Navini Island Resort Ltd, 28 January 1994

Roseman, J. – Managing Director, Tavarua Island Resort, 28 January 1994

Saukuru, J. – Assistant General Manager, Treasure Island Resort, 28 January 1994

Shaw, G.N. – Managing Director, Castaway Island, 27 January 1994

Sinclair-Hannocks, S. – Director, Department of Environment, Fiji Government, 3 February 1994

Singh, B. – Director, National Trust for Fiji, 1 February 1994

Singh, J. – Chief Executive, Fiji Hotel Association, 3 February 1994

Smith, D. – Managing Director, Musket Cove Resort, 27 January 1994

Steinocker, H. – Managing Director, Elegant Resorts Pty Ltd, 29 January 1994

Tuai, K. – Assistant Resort Manager, Beachcomber Island, 28 January 1994

Waqanisavou, T. – Estate Manager, Native Land Trust Board, 2 February 1994

Whiting, B. – Director of Marketing, Fiji Visitors Bureau, 2 February 1994

Whitton, R. – Director, Rosie the Travel Service, 4 February 1994

(ii) Informal interviews and/or discussions about the research were conducted with the following individuals

Anderson, A. – Former General Manager, Plantation Island Resort, currently Director, Lako Mai Resort (Malolo Island) (under development), 27 January 1994

Kabara, S. – Senior Research Officer, Reserve Bank of Fiji, 3 February 1994

Kudu, D. – Head of Planning, Tourism Council of the South Pacific, 3 February 1994

Rao, D. – Lecturer in Tourism, University of the South Pacific, 3 February 1994

Satyendra – Senior Officer, Fiji Government Department of Tourism, 3 February 1994

Thaggard, M. – Resort Manager, Musket Cove, 27 January 1994

Trusler, T. – Former General Manager, Suva Travelodge Hotel and President, Fiji Hotel Association, 4 July 1992

Walker, R. – Formerly Director, Pacific Ventures (Fiji) Ltd, 27 January 1994

Walker, T. – Formerly Resort Manager, Pacific Ventures (Fiji) Ltd (operators of Castaway Island), 27 January 1994

Whitton, T. – Managing Director, Rosie the Travel Service, 4 February 1994

Wong, P. – Former Resort Manager, Plantation Resort, December 1991

Government Policy and Legislation Unit, 10 November 1994

BIBLIOGRAPHY

Abali, A.Z. and Onder, D.E. (1990) 'The local architectural image in tourism', *Annals of Tourism Research* 17: 280–5.

Aboriginal and Torres Strait Islander Commission (ATSIC) (1994) *Draft National Aboriginal and Torres Strait Islander Tourism Strategy*, Canberra: AGPS.

Access Research (1989) *A Whole New Way to Fly: the Impact of Airline Deregulation, Computer Reservation Systems and Major Australian Travel Industry Developments*, Sydney: Australian Federation of Travel Agents.

—— (1990) *The Changing Australian Traveller*, Sydney: Australian Federation of Travel Agents.

—— (1992) *Delivering the Package*, Sydney: Australian Federation of Travel Agents.

Air Pacific (1993) *Annual Report 1992–1993*, Suva.

Amory, C. (1952) *The Last Resorts*, New York: Harper.

Anderssen, P. and Colberg, R. (1973) 'Multivariate analysis in travel research: a tool for travel package design and market segmentation', *Proceedings of the Travel Research Association, Fourth Annual Conference*, 225–40.

Anon. (1962) *Mackay and District 1862–1962*, Brisbane: Olive Ashworth Publishing Services.

Ash, J. (1974) 'The meaning of tan', *New Society* 1 August: 278–80.

Ashworth, G.J. (1991) 'Products, places and promotion: destination images in the analysis of the tourism industry' in M.T. Sinclair and M.J. Stabler (eds) *The Tourism Industry: An International Analysis*, 121–42, Wallingford: CAB International.

Ashworth, G.J. and Goodall, B. (eds) (1990) *Marketing Tourism Places*, London: Routledge.

Asian Hotelier (1994) March: 10.

ASMAL (1992) *The Australian Travel Market for Fiji*, prepared for the Fiji Ministry of Tourism, Auckland.

Australian Airlines (1992) *Annual Report 1991–2*, Melbourne.

Australian Bureau of Statistics (1992) *Short-Term Overseas Departures by Australian Residents*, Canberra: Australia Government Publishing Service.

—— (1993) *Queensland Pocket Year Book*, ABS Queensland Office.

—— (1995) *Overseas Arrivals and Departures*, Canberra: Australia Government Publishing Service.

Australian Concise Oxford Dictionary (1987) Melbourne: Oxford University Press.

Australian Hotelier (1994) 'High expectations', March: 6–7.

Ayala, H. (1991a) 'Resort hotel landscape as an international megatrend', *Annals of Tourism Research* 18(4): 568–87.

BIBLIOGRAPHY

—— (1991b) 'Resort landscape systems: a design management solution', *Tourism Management* December: 280–90.

—— (1993) 'Mexican resorts: A blueprint with an expiration date', *The Cornell Hotel and Restaurant Administration Quarterly* June: 34–42.

Bachelard, P. (1957) *La Poétique de l'Espace*, Paris: PUF.

Baines, G.B.K. (1977) 'The environmental demands of tourism in coastal Fiji' in J.H. Winslow (ed.) *The Melanesian Environment*, 9th Waigani Seminar, Canberra: UPNG and ANU.

Baker, M.J. (1985) *Marketing: an Introductory Text* 3rd edn, London: Macmillan.

Ballantyne, R.M. (1858) (1954 edn) *The Coral Island*, London: Dent.

Banfield, E.J. (1908) *The Confessions of a Beachcomber*, London: T.F. Unwin.

—— (1911) *My Tropic Isle*, London: T.F. Unwin.

—— (1918) *Tropic Days*, Sydney: Eagle Press.

—— (1925) *Last Leaves from Dunk Island*, Sydney: Eagle Press.

Barbaza, Y. (1970) 'Trois types d'intervention du tourisme dans l'organisation de l'éspace littoral', *Annales de Geographie* 434: 446–69.

Barbier, B. (1978) 'Ski et stations de sport d'hiver dans le monde', *Wiener Geographische Schriften* 51/52: 130–46.

Bard, S. (1989) 'Hyatt programs entertain kids', *Hotel and Motel Management* 204(1): 3.

Barker, B.C. (1992) *An Assessment of the Cultural Heritage Values on the Mainland Coast of the Whitsunday Region Extending from George Point to Repulse Bay*, published report to Queensland Department of Environment and Heritage.

Barnes, J. (ed) (1992) *Report on a Five-Year Funding Strategy for the Barrier Reef Marine Park*, unpublished report to the GBRMPA.

Barr, T. (1990) *No Swank Here? The Development of the Whitsundays as a Tourist Destination to the Early 1970s*, Studies in North Queensland History No. 15, Townsville: James Cook University.

Barrett, J.A. *The Seaside Resort Towns of England and Wales*, unpublished PhD thesis, University of London.

Barthes, R. (1957) *Mythologies*, Paris: Editions du Seuil.

Basuttin-Windsor, V. (1982) *Island that We Knew*, Mackay: V. Bassutin-Windsor.

Baud-Bovy, M. and Lawson, F. (1977) *Tourism and Recreation Development*, London: Architectural Press.

Baudrillard, J. (1983) *Simulations*, New York: Semiotext.

Bauer, E. (1986) *Pioneering a Tourist Island: Ernie Bauer's Story of South Molle Island*, Bowen: Bowen Independent.

Bayliss-Smith, T., Bedford, R., Brookfield, H. and Latham, M. (1988) *Islands, Islanders and the World: the Colonial and Post-Colonial Experience of Eastern Fiji*, Cambridge: Cambridge University Press.

Beaglehole, J.C. (1966) *The Exploration of the Pacific* 3rd edn, London: Adam & Charles Black.

—— (1974) *The Life of Captain James Cook*, London: Adam & Charles Black.

Beaver, A. (1980) 'Large scale inclusive tour operation', in A. Beaver, *Mind Your Own Travel Business*, 361–442, Edgware: Beaver Travel.

Becke, L. (1894) *By Reef and Palm*, London: T. Fisher Unwin.

Berger, A.A. (1991) *Media Analysis Techniques*, Beverly Hills: Sage.

Berger, J. (1972) *Ways of Seeing*, Harmondsworth, Middlesex: BBC and Penguin.

Berger, P. and Luckman, T. (1967) *The Social Construction of Reality*, Garden City: Doubleday.

Bergin, A. (1993) *Aboriginal and Torres Strait Islander Interests in the Great Barrier Reef Marine Park*, Townsville: GBRMPA.

Berman, M. (1982) *All that is Solid Melts into Air: the Experience of Modernity*, New York: Simon & Schuster.

Berriane, M. (1978) 'Un type d'éspace touristique Marocain: le littoral Méditerranéan', *Revue de Géographie du Maroc* 29(2): 5–28.

Blackwell, J. (ed) (1988) *The Tourism and Hospitality Industry*, Chatswood: Australian-International Magazine Services.

Blackwell, J. and Stear, L. (eds) (1989) 'Trans-Australian Airlines "Take-7" Holidays', *Case Histories of Travel and Hospitality*, 236–44, Chatswood: Australian-International Magazine Services.

Blainey, G. (1976) *The Triumph of the Nomads: a History of Ancient Australia*, Melbourne: Macmillan.

Bloch, E. (1959) *Das Princip Hoffnung*, Frankfurt am Main: Suhrkamp Verlag.

Boers, H. and Bosch, M. (1994) *The Earth as a Holiday Resort: an Introduction to Tourism and the Environment*, Utrecht: SME.

Boldrewood, R. (1881) (1985 edn) *Robbery under Arms*, Ringwood: Penguin Books.

Boniface, P. and Fowler, P.J. (1993) *Heritage and Tourism in the Global Village*, London: Routledge.

Boorstin, D.J. (1964) *The Image: A Guide to Pseudo-Events in America*, New York: Harper & Row.

—— (1973) *The Americans: the Democratic Experience*, New York: Random House.

Bosselman, F.P. (1978) *In the Wake of the Tourist*, Washington DC: The Conservation Foundation.

Bourdieu, P. (1984) *Distinction. A Social Critique of the Judgement of Taste*, trans. R. Nice, London: Routledge and Kegan Paul.

Breinl, A. and Young, W.J. (1920) 'Tropical Australia and its settlement', *Annals of Tropical Medicine and Parasitology* 13: 351–412.

Britton, R.A. (1979) 'The image of the Third World in tourism marketing', *Annals of Tourism Research* 6: 318–29.

Britton, S.G. (1980) 'The spatial organisation of tourism in a neo-colonial economy: a Fiji case study', in *Pacific Viewpoint* 21(2): 144–65.

—— (1982) 'The political economy of tourism in the Third World', *Annals of Tourism Research* 9(3): 331–58.

—— (1983) *Tourism and Underdevelopment in Fiji*, Monograph No.13, Canberra: ANU Development Studies Centre.

Britton, S.G. and Clarke, W.C. (eds) (1987) *Ambiguous Alternatives: Tourism in Small Developing Countries*, Suva: USP.

Brookfield, H.C. and Hart, D. (1971) *Melanesia: a Geographical Interpretation of an Island World*, London: Methuen.

Brown, S. (1991) 'The day the United States Navy attacked the people of Malolo', *Islands* 1: 47–50.

Buber, M. (1966) *Paths in Utopia*, trans. R.F.C. Hull, Boston: Beacon Press.

Budowski, G. (1976) 'Tourism and conservation: conflict, coexistence or symbiosis', in *Environmental Conservation* 3(1): 27–31.

Bugden, G.F. (1992) 'Integrated resorts', paper presented at the *Law Council of Australia, General Practice Section Annual Conference*.

Bureau of Tourism Research (1990) *Tourism Trends in Australia*, Canberra.

—— (1994) *Domestic Tourism Monitor 1992/3*, Canberra.

Burkart, J. and Medlik, S. (1981) *Tourism Past, Present and Future*, London: Heinemann.

Burnet, L. (1963) *Villegiature et Tourisme sur les Côtes de France*, Paris: Hachette.

Butler, R. (1980) 'The concept of a tourist resort life cycle of evolution: implications for management of resources', *Canadian Geographer* 34(1): 5–12.

Bywater, M. (1990a) 'Airlines in the South Pacific', *Travel and Tourism Analyst* 1: 5–28.

—— (1990b) 'Spas and health resorts in the EC', *Travel and Tourism Analyst* 6: 52–67, London: Economist Intelligence Unit.

—— (1991) 'Prospects for Mediterranean beach resorts: an Italian case study', *Travel and Tourism Analyst* 5: 75–89, London: Economist Intelligence Unit.

Cabaniuk, S. (1992) *A Background Information Report on the Mamanuca Island Group*, Suva: IUCN Consultants, National Environment Management Project.

Calder, W. (1981) *Beyond the View: Our Changing Landscapes*, Melbourne: Inkata Press.

Cals, J., Esteban, J. and Teixidor, C. (1977) 'Les processes d'urbanisation touristique sur la Costa Brava', *Révue Géographique des Pyrénées et du Sud-Ouest* 48: 199–208.

Cameron McNamara (1987) *Cairns Region Joint Research Study*, Cairns: Cairns City and Mulgrave Shire Councils.

Carruthers, F. and Cant, S. (1994) 'Commonwealth hits States on park management deals', *The Weekend Australian* October 29–30: 12.

Cato, A.C. (1950) 'Malolo Island and Viseisei Village, Western Fiji', *Oceania* [a journal devoted to study of the native peoples of Australia, New Guinea and the Islands of the South Pacific] 23(2).

Cazes, G. (1976) *Le Tiers Monde Vu Par Les Publicités Touristiques: Une Image Géographique Mystifiante*, Les Cahiers du Tourisme C–33, Aix-en-Provence: CHET.

—— (1984) *Tourisme Enclave, Tourisme Intègre: Le Grand Débat de L'Aménagement Touristique dans les Pays en Développement*, Les Cahiers du Tourisme C–101, Aix-en-Provence: CHET.

—— (1987) *L'Isle Tropicale, Figure Emblèmatique du Tourisme International*, Les Cahiers du Tourisme, Series C, No.112, Aix-en-Provence: CHET.

Ceballos-Lascurain, H. (1991) 'Tourism, ecotourism and protected areas', *Parks* 2(3), November: 31–5.

Central Planning Office (1966) *Fiji Development Plan 1966–70: Development Planning Review*, Suva: Central Planning Office, Legislative Council of Fiji, Council Paper No. 11 of 1966.

Chamberlain, R.N. (1983) 'Scheveningen, the Hague: the revitalisation of a declining holiday resort', *Developing Tourism*, 25–33, London: PRTC Education and Research Services.

Chambers, M. (1990) *Guidelines for the Integration of Tourism Development and Environment Protection in the South Pacific*, Suva: TCSP.

Charlton, C.B. (1992) 'Integrated resorts – Queensland's statutory schemes', paper presented at *Law Council of Australia General Practice Section Annual Conference*.

Checchi & Co. (1961) *The Future of Tourism in the Pacific and Far East*, Washington DC.

Christaller, W. (1963) 'Some considerations of tourism location in Europe: the peripheral regions in underdeveloped countries' recreation areas', *Regional Science Association Papers* 12: 95–105.

Cilento, R.W. (1923) *Climatic conditions in North Queensland as they affect the health and virility of the people*, Brisbane: Government Printer.

Cilento, R.W. and Lack, C. (1959) *Triumph in the Tropics: an Historical Sketch of Queensland*, Brisbane: Smith & Paterson.

Claringbould, R., Deakin, J. and Foster, P. (1984) *Data Review of Reef Related Tourism 1946–1980*, Townsville: GBRMPA.

Club Med (1994) *Club Med Lindeman Island, Whitsundays Queensland*, brochure valid May 1994 to 1995.

Cognat, B. (1973) *La Montagne Colonisée*, Paris: PUF.

Cohen, E. (1972) 'Towards a sociology of international tourism', *Social Research* 39: 164–82.

—— (1982) *The Pacific Islands from Utopia Myth to Consumer Product: the Disenchantment of Paradise,* Cahiers du Tourisme B–27, Aix-en-Provence: CHET.

Colfelt (1985) *100 Magic Miles of the Great Barrier Reef: the Whitsunday Islands* 2nd edn, Sydney: David Colfelt and Windward Publications.

Colliers Jardine (1994) *Australian Hotel and Tourism Property Market Report,* Melbourne: Colliers Jardine.

Collins, C.O. (1979) 'Site and situation strategy in tourism planning: a Mexican case study', *Annals of Tourism Research* 6(3): 351–66.

Commonwealth Department of Tourism (1994a) *National Ecotourism Strategy,* Canberra: AGPS.

—— (1994b) *National Rural Tourism Strategy,* Canberra: AGPS.

Commonwealth Department of Transport and Communications, *Scheduled Domestic Air Transport,* various reports, Canberra: AGPS.

—— *Scheduled International Air Transport,* various reports, Canberra: AGPS.

Coopers and Lybrand (1989) *Tourism Masterplan for Fiji,* Suva: report prepared for the Fiji Ministry of Tourism.

Coulter, J.W. (1946) 'Impact of war on the South Sea Islands', *Geographical Review* 36 (3): 409–19.

Courbis, R. (1984) 'Les conséquences économiques and sociales de l'aménagement touristique: l'impact des investissements réalisés dans le cas du Languedoc-Roussillon', *World Travel* 180: 29, 33.

Cowell, D. (1984) *The Marketing of Services,* London: Heinemann.

Cox, P. (1985) 'The architecture and non-architecture of tourist developments', *Tourist Developments in Australia,* 46–51, Red Hill: Royal Australian Institute of Architects Education Division.

Craik, J. (1987a) 'A crown of thorns in paradise: tourism on Queensland's Great Barrier Reef', in M. Bouquet and M. Winter (eds), *Who From their Labours Rest? Conflict and Practice in Rural Tourism,* 135–58, Aldershot: Avebury.

—— (1987b) 'From cows to croissants: Creating communities around leisure and pleasure', *Social Alternatives* Vol. 6 (3): 21–7.

—— (1988) *Tourism Down Under: Tourism Policies in the Tropics,* Cultural Policy Studies, Occasional Paper No.2, Brisbane: Griffith University.

—— (1991) *Resorting to Tourism: Cultural Policies for Tourist Development in Australia,* Sydney: Allen & Unwin.

Crocombe, R. (1987) *The South Pacific: An Introduction* 4th edn, Auckland: Longman Paul.

Crompton, J. (1979) 'Motivations for pleasure travel', *Annals of Tourism Research* 4(4): 184–94.

Cumin, G. (1970) 'Les stations integrées', *Urbanisme* 116: 50–3.

Curran, P. (1978) *Principles and Procedures of Tour Management,* Boston: CBI.

Daniels, S. and Cosgrove, D. (1988) 'Introduction: iconography and landscape', in D. Cosgrove and S. Daniels (eds), *The Iconography of Landscape,* 1–10, Cambridge: Cambridge University Press.

Dann, G.M.S. and Potter, R.B. (1994) *Tourism and Postmodernity in a Caribbean Setting,* Les Cahiers du Tourisme C–185, Aix-en-Provence: CHET.

Davidson, L. (1994) 'Landscape with words: writing about landscape', *Overland* 134 Autumn: 6–10.

Dawson, K.C.A. (1987) *Malolo Island, Fiji: an Initial Coastal Archaeological Survey 1986–87,* Suva, Government Press.

Day, P. (1994) *Draft Recreation and Sport Plan, Cannonvale Area*, Whitsunday: report prepared for the Whitsunday Shire Council.

Dean, J. and Judd, B. (eds) (1985) *Tourism Developments in Australia*, Red Hill: Royal Australian Institute of Architects Education Division.

Debord, G. (1983) *Society of the Spectacle*, Detroit: Black and Red.

Defert, P. (1988) *Problématique du Tourisme Insulaire*, Les Cahiers du Tourisme C–58, Aix-en-Provence: CHET.

Defoe, D. (1719) (1993 edn) *Robinson Crusoe*, London: Penguin Books.

Department of Tourism, Sport and Racing (DTSR) (1994) *Queensland Ecotourism Strategy Discussion Paper*, Brisbane: QGPS.

Derrick, R.A. (1951) *The Fiji Islands: A Geographical Handbook*, Suva, Government Press.

Dickinson (1982) 'The three billion boom in holiday resorts', *National Times* June 13–19: 39.

Dieke, P.U.C. (1993) 'Cross-national comparison of tourism development', *Journal of Tourism Studies* 4, (1): 2–18.

Dobinson, J. (1987) *Integrated Resort Development. A Guide for Local Authorities, Planners and Developers*, Brisbane: Local Government Association of Queensland.

Dorfmann, M. (1983) 'Régions de montagne: de la dépendance à la autodéveloppement', *Revue de Géographie Alpine*, 71(1): 5–34.

Dowling, J. (1993) 'Barrier Reef resorts boom for Qantas after airline merger', *Australian Financial Review* 21 December: 13.

—— (1994) 'Hamilton Island looking for new markets', *Australian Financial Review* 20 April: 40.

Doyle, T. (1989) 'Lindeman Island: environmental politics in Queensland', *Journal of the Royal Historical Society of Queensland* 13(12): 462–72.

Dufour, R. (1978) *Des Mythes du Loisir/Tourisme Weekend: Aliénation ou Libération?*, Les Cahiers du Tourisme C–47, Aix-en-Provence: CHET.

Dumas, D. (1975) 'Un type d'urbanisation touristique littorale: la Manga del Mar Menor (Espagne)', *Travaux de L'Institut de Géographie de Rheims*, 23–4, 89–96.

—— (1976) 'L'urbanisation touristique du littoral de la Costa Blanca (Espagne)', *Cahiers Nantais* 13: 43–50.

Dumazedier, J. (1967) *Towards a Society of Leisure*, trans. S.E. McLure, New York: Free Press.

Dunning, J.H. and McQueen, M. (1982a) 'Multinationals in the international hotel industry', *Annals of Tourism Research* 9(1): 69–90.

—— (1982b) *Transnational Corporations in International Tourism*, New York: United Nations Centre for Transnational Studies.

Dyer, G. (1982) *Advertising as Communication*, London: Methuen.

Eco, U. (1986) *Travels in Hyperreality*, London: Picador.

Economist Intelligence Unit (1978) *Air Inclusive Tour Marketing: the Retail Distribution Channels in the UK and West Germany*, London: Economist Publications.

—— (1987) 'Fiji', *International Tourism Reports*, London.

Ecotourism Association of Australia (1992) *Newsletter* 1(2).

Eliade, M. (1952) *Images et Symboles*, Paris: Tel-Gallimard.

—— (1968) *Myth, Dreams and Mysteries*, London: Collins.

—— (1969) 'Paradise and Utopia: mythical geography and eschatology', in M. Eliade (ed.) *The Quest*, 88–111, Chicago & London: Chicago University Press.

—— (1971) *The Myth of Eternal Return*, Princeton: Princeton University Press.

Elliott, T. (1995) 'Five blockbuster marketing ideas', *Australian Professional Marketing* March: 10–15.

260

Elton, M.A. (1984) 'UK tour operators and retail travel agents: ABTA and the public interest', *Tourism Management* 5(3): 223–8.

England, R. (1980) 'Architecture for tourists', *International Social Science Journal* 32(1): 44–5.

Farrell, B.H. (1982) *Hawaii, the Legend that Sells*, University Press of Hawaii.

Farwell, T.A. (1970) 'Resort planning and development', *Cornell Hotel and Restaurant Administration Quarterly* February: 34–7.

Feifer, M. (1985) *Going Places: Tourism in History from Imperial Rome to the Present*, New York: Stein & Day.

Figgis, P. (1992) *Eco-tourism: Special Interest or Major Direction?*, 2nd rev., July.

Fiji Government Department of Tourism (1992) *General Information on Tourism in Fiji: Its Past and Future and Impact on the Economy and Society*, Suva: Fiji Government Department of Tourism.

Fiji Hotels Association (1993) *Unpublished Minutes of a Meeting Between Tour Operators, the Accommodation Sector and Airlines*, Sydney.

Fiji Visitors Bureau (FVB) (1993a) *1994 Marketing Plan*, Suva, FVB.

—— (1993b) *A Statistical Report on Visitor Arrivals into Fiji: Calendar Year 1992*, Suva: ASMAL.

—— (1993c) *International Visitor Survey programme 1992: Visit Characteristics Survey*, Auckland, Suva: ASMAL.

—— (1995) *Fiji International Visitor Survey*, Suva: ASMAL.

Fitzgerald, R. (1982) *From the Dreaming to 1915: a History of Queensland*, Brisbane: Queensland University Press.

—— (1984) *A History of Queensland From 1915 to the 1980s*, Brisbane: QUP.

Fleetwood, S. (1993) 'Standard classification of visitor accommodation' *Building a Research Base in Tourism: Proceedings of the National Conference on Tourism Research*, 165–72, University of Sydney.

Flood, J., Maher, P. and Roy, J. (1991) *The Determinants of Internal Migration in Australia*, Melbourne: CSIRO Division of Building, Construction and Engineering.

Forer, P.C. and Pearce, D.G. (1984) 'Spatial patterns of package tourism in New Zealand', *New Zealand Geographer* 40(1): 34–42.

Foss, P.(ed) (1988) *Islands in the Stream: Myths of Place in Australian Culture*, Sydney: Pluto Press.

Foster, G.M. (1986) 'South Sea cruise: a case study of a short-lived society', *Annals of Tourism Research* 13: 215–38.

Frappat, P. (1979) *Le Mythe Blessé*, Paris: PUF.

Freitag, T.G. (1994) 'Enclave tourism development: for whom the benefits roll?', *Annals of Tourism Research* 21(3): 538–54.

Frings, J.W. (Undated) *My Island of Dreams*, London: Newnes.

Frodey, C. and O'Hara, J. (1992) 'Are European "quality service" models applicable to resorts in the South Pacific?', *Journal of Pacific Studies* 16: 90–107.

Fukuyama, F. (1989) 'The End of History?', *The National Interest* Summer: 1–18.

Gamage, A., King, B.E.M. and Wise, B. (1989) *Conducting Tourism Feasibility Studies*, Melbourne: Victoria University of Technology.

Gartner, W.C. (1986) 'Temporal influences on image change', *Annals of Tourism Research* 13: 635–44.

—— (1993) 'Image formation process', *Journal of Travel and Tourism Marketing* 2 (2/3): 191–216.

Gatty, B. (1991) 'Child care issues gain prominence in congress', *Hotel and Motel Management* 206(1): 16.

Gazzard, D. (1985) 'Dunk Island Resort', in J. Dean and B. Judd (eds), *Tourist Developments in Australia*, 67–8, Royal Australian Institute of Architects.

Gee, C. Y., (1988) *Resort Development and Management* 2nd edn, East Lansing, Michigan: Educational Institute of the American Hotel and Motel Association.

Gennep, A.V. (1960) *The Rites of Passage*, trans. M.V. Vizedom and G.L. Caffee, London: Routledge & Kegan Paul.

Ghent, A. (1992) 'A last resort to save an island', *Age* 16 May: 17.

Giamatti, A.B. (1966) *The Earthly Paradise and the Renaissance Epic*, Princeton, NJ: Princeton University Press.

Giesz, L. (1968) 'Kitsch man as tourist', in G. Dorfles (ed.), *Kitsch: the World of Bad Taste*, 17–26, New York: Bell Publications.

Gilmour, A.J., Cunningham, J. and Lamond, D. (1991) *Day to Day Management of the Great Barrier Reef Marine Park: the Role of the Marine Park Staff*, Macquarie University report to the GBRMPA.

Gladstone, C. (1995) 'QF sales to new heights', *Traveltrade* 8 February: 2.

Glaister, K. (1991) 'International success: Company strategy and national advantage', *European Management Journal* 9(3): 334–8.

Go, F. (1989) 'Resorts in North America: problems and prospects', *Travel and Tourism Analyst*, 19–36, London: Economist Intelligence Unit.

Goffman, E. (1959) *The Presentation of Self in Everyday Life*, Harmondsworth: Penguin.

Golding, W. (1954) *Lord of the Flies*, London: Faber.

Goodall, B., and Bergsma, J. (1990) 'Destinations – as marketed in tour operators' brochures', in G. Ashworth and B. Goodall (eds), *Marketing Tourism Places*, 170–92, London: Routledge.

Goodall, B., Radburn, M. and Stabler, M. (1988) 'Market opportunity sets for tourism', in *Geographical Paper 100: Tourism Series*, Reading: Department of Geography, University of Reading.

Goodrich, J.N. (1977) 'Differences in perceived similarity of tourism regions: a spatial analysis', *Journal of Travel Research* 16(1): 10–13.

—— (1978) 'The relationship between preferences for and perceptions of vacation destinations: application of a choice model', *Journal of Travel Research* 17 (Fall): 8–13.

Gordon, W. and Langmaid, R. (1988) *Qualitative Market Research: A Practitioner's and User's Guide*, Aldershot: Gower.

Gormsen, E. (1981) 'The spatio-temporal development of international tourism: attempt at a centre–periphery model', *La Consommation d'Espace par le Tourisme et sa Préservation*, 150–70, Aix-en-Provence: CHET.

Gottlieb, A. (1982) 'Americans' vacations', *Annals of Tourism Research* 9: 165–87.

Government of Queensland, Office of the Co-ordinator General (1994) *Whitsunday Tourism Strategy*, draft April, Brisbane.

Great Barrier Reef Marine Park Authority (GBRMPA) (1981) *Nomination of the Great Barrier Reef by the Commonwealth of Australia for Inclusion on the World Heritage List*, Townsville: GBRMPA.

—— (1993) *Annual Report 1992–93*, Townsville: GBRMPA.

Great Barrier Reef Marine Park Authority and the Queensland Department of Environment and Heritage (1993) *A 25 Year Strategy Plan for the Great Barrier Reef World Heritage Area*, Townsville: GBRMPA.

Green, G. and Lal, P. (1991) *Charging Users of the Great Barrier Reef Marine Park*, Townsville: GBRMPA.

Grey, P. (1970) *International Travel – International Trade*, Lexington: Grid Publishing.

Grunthal, A. (1936) *Probeme der Fremdenverkehrs: Geographie*, Schriftreihe des Forschungs Institut für den Fremdenverkehr FRG Berlin Heft.9.

Guerin, J-P. (1984) *L'Aménagement de la Montagne en France: Politique, Discours et Productions d'Espace dans les Alpes du Nord*, Paris: Ophyrs.

Guitart, C. (1982) 'UK charter flight package holidays to the Mediterranean, 1970–1978', *Tourism Management* March: 16–39.

Gunn, C. (1988a) *Tourism Planning* 2nd edn, New York: Taylor & Francis.

—— (1988b) *Vacationscape: Designing Tourist Regions* 2nd edn, New York: Van Nostrand Reinhold.

Gyte, D.M. and Phelps, A. (1989) 'Patterns of destination repeat business: British tourists in Mallorca, Spain', *Journal of Travel Research* 28 (Summer): 24–8.

Haigh, R. (1993) *Keeping Australians at Home: Tourism Import Replacement Analysis*, Occasional Paper 17, Canberra: Bureau of Tourism Research.

Hair, J.F. (1992) *Multivariate Data Analysis with Readings* 3rd edn, New York: Macmillan.

Hall, C.M. (1991) *Introduction to Tourism in Australia: Impacts, Planning and Development*, Melbourne: Longman Cheshire.

—— (1994a) 'The closer economic relationship between Australia and New Zealand: implications for travel and tourism', *Journal of Travel and Tourism Marketing* 3(1): 123–32.

—— (1994b) *Tourism in the Pacific Rim: Development, Impacts and Markets*, Melbourne: Longman Cheshire.

Hanlon, P. (1989) 'Hub operations and airline competition', *Tourism Management* 10, No.2: 111–24.

Harris, Kerr, Foster and Company (1965) *Report on a Study on the Travel and Tourism Industry in Fiji*, Council Paper 32, Suva: Legislative Council of Fiji.

Harrison, D. (ed.) (1992) *Tourism and the Less Developed Countries*, London: Bellhaven.

Hawkins, D., Shafer, E. and Rovelstad, J. (1980) *Tourism Planning and Development Issues*, Washington DC: George Washington University.

Hemmeter, C. (1988) 'Resort development', *Proceedings of the Pacific Asia Travel Association Annual Conference*, Melbourne.

Herridge, N. (1990) 'The holiday market after deregulation: a wholesaler's perspective', in B.E.M. King (ed.), *Domestic Air-Inclusive Holidays after Deregulation: Challenges and Opportunities for the Retailer*, 1–4, Melbourne: Victoria University of Technology.

Heywood, P. 'Beauty, truth and landscape: trends in landscape and leisure', *Queensland Planner* 28 (4): 2–7.

Hirsch, F. (1978) *Social Limits to Growth*, London: Kegan Paul.

Hirschman, E. and Holbrook, M. (1982) 'Hedonic consumption: Emerging concepts, methods and propositions', *Journal of Marketing* 46: 92–101.

Hoivik, T. and Heiberg, T. (1980) 'Centre-periphery tourism and self-reliance', *International Social Science Journal* 32(1): 69–98.

Hollinshead, K. (ed.) (1985) *Tourist Resort Development: Markets, Plans and Impacts*, Conference Proceedings, Sydney.

Holloway, J.C. and Plant, R.V. (1992) *Marketing for Tourism* 2nd edn, London: Pitman.

Holmes, J. H. (1985) 'Underlying themes in the geography of Queensland', in J.H. Holmes (ed.), *Queensland: A geographical interpretation*, Brisbane: Queensland Geographical Journal, 4th Series.

Hooper, P. (1994) *Evaluating Strategies for Packaging Travel*, Sydney: University of Sydney Institute of Transport Studies Working Paper.

Horwath & Horwath (1988) *The Hotel Development Process*, Sydney, Australia.

Hotels (1993) 'Give your hotels an environmental checkup', October: 59–60.

Howard, J. and Sheth, J.N. (1969) *Theory of Buyer Behaviour*, New York: John Wiley & Sons.

Huizinga, J. (1950) *Homo Ludens: A Study of the Play Element in Culture*, Boston: Beacon.

Hundloe, T., Neumann, R. and Halliburton, M. (1989) *Great Barrier Reef Tourism*, Brisbane: Institute of Applied Environmental Research, Griffith University.

Hunt, J.D. (1975) 'Image as a factor in tourism development', *Journal of Travel Research* 13(3): 1–7.

I.D.P. Interdata Pty Ltd (1988, 1989, 1990, 1991) (four editions) *The Interdata Leisure and Tourism Handbook: the Leisure and Tourism Development and Investment Handbook of Australia, New Zealand and the South West Pacific*, Sydney: IDP Publishers.

Innskeep, E. (1994) *National and Regional Tourism Planning. Methodologies and Case Studies*, London: Routledge.

Inside Tourism (1988) 'Williams warns investors on resorts', 31 October: 13.

—— (1991a) 'Hamilton Island to float now case is settled', 18 November: 3.

—— (1991b) 'Hayman turns to agents to fill space', 12 August: 6.

International Air Transport Association (1984) *Asia-Pacific Passenger Traffic Forecast: Travel Demand 1985–2000*, San Francisco: Pacific Asia Travel Association.

IUCN, The World Conservation Union (1992) *Environment: Fiji, The National State of the Environment Report*, Gland, Switzerland: IUCN.

—— (1993) *The National Environment Strategy: Fiji*, Suva: Report prepared for the Fiji Ministry of Housing and Urban Development.

Jason (1994) 'The end of the road for the banks in travel', *Traveltrade* 21 December: 6.

Jenkins, C.L. (1980) 'Tourism policies in developing countries: a critique', *International Journal of Tourism Management*, 9 (1): 22–9.

—— (1982) 'The effects of scale on tourism in developing countries', *Annals of Tourism Research* 9: 229–49.

Johnson, I. (1985) 'Issues in coastal resort centres in Queensland', *Australian Urban Studies* 13,1: 7–9.

Johnston, E. (1985) 'End of an era for Hayman Island', *Weekend Australian* 20–21 July: 32.

Jung, C.G. (1959) *Four Archetypes: Mother, Rebirth, Spirit, Trickster*, trans. R.F.C. Hull, Princeton NJ: Princeton University Press.

Juvik, J. and Singh, B. (1989) 'Conservation of the Fijian Crested Iguana: a progress report', presented at *Fourth South Pacific Conference on Nature Conservation and Protected Areas*, Port Vila, Vanuatu.

Kaiser, C. Jnr. and Helber, L. (1978) *Tourism Planning and Development*, Boston: CBI.

Keith-Reid, R. (1994a) 'Room please: the call that threatens slow growing Fiji', *Islands Business Pacific Annual State of Tourism Report*, 27–9.

——(1994b) 'The South Seas man', *Islands* 1: 18–24.

Kennedy, F. (1995) 'Great White Hope fails to deliver on a black promise', *Australian* 19 January: 7.

Kermath, B.M. and Thomas, R.N. (1992) 'Spatial dynamics of resorts: Sosua, Dominican Republic', *Annals of Tourism Research* 19: 173–90.

King, B.E.M. (ed.) (1990) *Domestic Air-Inclusive Holidays after Deregulation. Challenges and Opportunities for the Retailer*, Melbourne: seminar proceedings published by Footscray Institute of Technology.

—— (1991) 'Tour operators and the air-inclusive tour industry in Australia', *Travel and Tourism Analyst* 3: 66–87, London: Economist Intelligence Unit.

—— (1992a) 'The Australian and New Zealand long haul travel market', *Travel and Tourism Analyst* 6: 47–70.

—— (1992b) 'Cultural tourism and its potential for Fiji', *Journal of Pacific Studies* 16(1): 74–89.

—— (1994) 'Bringing out the authentic in Australian hospitality products for the international tourist: a service management approach', *Australian Journal of Hospitality Management* 1(1): 1–8.

King, B.E.M. and Hyde, G. (1989a) *Tourism Marketing in Australia*, Melbourne: Hospitality Press.

—— (1989b) 'Resorts', *Tourism Marketing in Australia*, 197–227, Melbourne: Hospitality Press.

King, B.E.M. and McVey, M. (1994) 'Fiji', *International Tourism Reports* 2: 5–22, London: Economist Intelligence Unit.

—— (1996a) 'Australia outbound', *Travel and Tourism Analyst* 3: 21–39, London: Corporate Intelligence.

—— (1996b) 'Resorts in Asia', *Travel and Tourism Analyst* 4: 35–50, London: Corporate Intelligence.

King, B.E.M. and Weaver, S. (1993) 'The impact of the environment on the Fiji tourism industry: a study of industry attitudes', *Journal of Sustainable Tourism* 1(2): 97–111.

King, B.E.M. and Whitelaw, P. (1992) 'Resorts in Australian tourism: a recipe for confusion?', *Journal of Tourism Studies* 4, No.1.

King, B.E.M., Pizam, A. and Milman, A. (1993) 'Social impacts of tourism: host perceptions', *Annals of Tourism Research* 20: 650–65.

King, J. (1984) 'The air traffic market and tourism: some thoughts on the South Pacific', in C.C. Kissling (ed.), *Transport and Communications for Pacific Microstates*, 113–26, Suva: Issues in Organisation and Management, Institute of Pacific Studies.

King, M. (1981) 'Disneyland and Walt Disney World: transitional values in futuristic form', *Journal of Popular Culture* 15: 116–40.

Knafou, R. (1978) 'Les stations integrées de sports d'hiver des Alpes Françaises', *L'Espace Géographique* 8(3): 173–80.

Knight, J.B. and Salter, C.A. (eds) (1987) *Foodservice Standards in Resorts*, New York: CBI/Van Nostrand Reinhold.

Kojeve, A. (1969) *Introduction to the Reading of Hegel*, Allan Bloom and Raymond Queneau (eds), trans. James Nichols, New York: Basic Books.

Kong, C.C. (1994) 'Towards a multilateral air transport regime', *Viewpoint* 1(1): 22–30, World Travel and Tourism Council.

Kotler, P. (1991) *Marketing Management: Analysis, Planning and Control* 5th edn, London: Prentice-Hall International.

Krippendorf, J. (1987) *The Holidaymakers: Understanding the Impact of Leisure and Travel*, trans. V. Andrassy, Oxford: Heinemann.

Lamond, H.G. (1948) 'An island holiday', *Cummins and Campbell's Monthly Magazine* March: 13–15.

Lane, B. (1994) 'What is rural tourism?' *Journal of Sustainable Tourism* 2 (1 and 2): 7–21.

Lang, J. (1990) 'Cultural and social futures for Australian cities', in D. Wilmoth (ed.), *Urban Futures: Towards an Agenda for Australian Cities*, Canberra: DITAC.

Lavarack, P.W. (1985) *Draft Whitsunday Management Plan* (unpublished), Queensland Department of Environment and Heritage internal report.

Lavery, P. (1989) 'Indoor resorts in the EC', *Travel and Tourism Analyst* 1: 52–68.

Lawson, F. (1976) *Hotels, Motels and Condominiums: Design, Planning and Maintenance*, Boston: CBI.

Layton, R. (ed.) (1979) 'Parkview (Keppel) Pty Ltd, The Campaign Palace: we must be doing something right on Great Keppel Island', *Australian Marketing Projects: Hoover Award for Marketing 1978*, Sydney.

Lea, J. (1988) *Tourism and Development in the Third World*, London: Routledge.

Lea, J. and Small, J. (1988) 'Cyclones, riots and coups: tourist industry responses in the South Pacific', in B. Faulkner and M. Fagence (eds), *Frontiers in Australian Tourism: the Search for New Perspectives in Policy Development and Research*, Canberra: Bureau of Tourism Research.

Leers, T.J. (1983) 'From salvation to self-realisation: advertising and the therapeutic roots of the consumer culture 1880–1930', in R.W. Fox and T.J. Leers (eds), *The Culture of Consumption: Critical Essays in American History 1880–1930*, New York: Pantheon.

Lefebvre, H. (1984) *Everyday Life in the Modern World*, trans. S. Rabinovitch, New Brunswick NJ: Transaction Books.

Leiper, N. (1979) 'The framework of tourism: towards a definition of tourism, tourist and the tourist industry', *Annals of Tourism Research* 6(4): 390–407.

—— (1980) *An Interdisciplinary Study of Australian Tourism: its Scope, Characteristics and Consequences, with Particular Reference to Governmental Policies Since 1965* (unpublished Master of General Studies Thesis), University of New South Wales, Sydney.

—— (1981) 'Towards a cohesive curriculum in tourism studies', *Annals of Tourism Research* 8(1): 69–84.

—— (1990a) *Tourism Systems. An Interdisciplinary Perspective*, Occasional Paper 2, Palmerston North: Massey University Department of Management Systems.

—— (1990b) 'The partial industrialisation of tourism systems', *Annals of Tourism Research* 17(4): 600–5.

Lever, A. (1987) 'Spanish tourism migrants: the case of Lloret del Mar', *Annals of Tourism Research* 14(4): 449–70.

Levi-Strauss, C. (1970) *The Raw and the Cooked*, New York: Harper & Row.

Lewis, S.L. and Brissett, D. (1981) 'Paradise on demand', *Society* 18(5): 85–90.

Leymore, V.L. (1975) *Hidden Myth*, New York: Basic Books.

Littlefield, J.E. and Kirkpatrick, C.A. (1970) *Advertising: Mass Communication in Marketing*, Boston: Houghton Mifflin.

Lock, A.C.C. (1955) *Destination Barrier Reef*, Melbourne: Georgian House.

—— (1956) *Tropical Tapestry*, Melbourne: Georgian House.

Lodge, D. (1991) *Paradise News*, London: Secker and Warburg.

London, J. (1911) *South Sea Tales*, New York: Macmillan.

Lovelock, C.H. (1984) *Services Marketing: Text, Cases and Readings*, New Jersey: Prentice-Hall.

Lowenthal, D. (1981) 'Conclusion: dilemmas of preservation', in M. Binney and D. Lowenthal (eds), *Our Past Before Us: Why do we Save it?*, 213–37. London: Temple Smith.

Lynch, K. (1960) *The Image of the City*, Cambridge Mass.: MIT Press and Harvard University Press.

Mabogunje, A.L. (1980) *The Development Process: A Spatial Perspective*, London: Hutchinson.

McCalman, J. (1994) 'Suburbia from the sandpit', *Meanjin Quarterly* 53(3): 548–53.

MacCannell, D. (1976) *The Tourist*, New York: Schocken Books.

—— (1992) *Empty Meeting Grounds: the Tourist Papers*, London: Routledge.

BIBLIOGRAPHY

McCracken, G. (1982) *Culture and Consumption: New Approaches to the Symbolic Character of Consumer Goods and Activities*, Bloomingdale: Indiana University Press.

MacDermott, K. and O'Meara, M. (1993) 'Citibank decides to float Hamilton Island', *Australian Financial Review* 16 September: 1, 6.

McDermott Miller (1993) *National Parks in Fiji: Economic Assessment*, Wellington: prepared for the Maruia Society and the Native Land Trust Board of Fiji.

McDonald, M.H.B. (1984) *Marketing Plans*, London: Heinemann.

MacDonnell, I. (1994) *Leisure Travel to Fiji and Indonesia from Australia 1982 to 1992: Some Factors Underlying Changes in Market Share*, MA thesis, University of Technology, Sydney.

McIntyre, A. (1992) *An Analysis of Fiji's Tourist Accommodation Structure and Room Constraints 1991–96*, Auckland: ASMAL.

Main, K. (1989) *Airline Development in the South Pacific: A Turning Point*, M.Bus. Thesis, Victoria University of Technology, Melbourne.

Marshman, I. (1992) 'Hamilton Island show is not over until the fat lady sings: Williams', *Traveltrade* 3 June: 10.

—— (1994) 'Ansett next for Bali', *Traveltrade* 21 September 1994: 1.

—— (1995) 'NZ reaps rewards from move into short breaks', *Traveltrade* 8 March: 8.

Mattelart, A. (1991) *Advertising International: the Privatisation of Public Space*, trans. M. Chanan, London: Routledge.

Meinig, D. (1979) 'The beholding eye', in D.W. Meinig (ed.), *The Interpretation of Ordinary Landscapes. Geographic Essays*, 33–48. New York: Oxford University Press.

Mercer, P.M. (1974) 'Pacific Islanders in Colonial Queensland 1863–1906', in *Lectures on North Queensland History*, 101–20, Townsville: Department of History, James Cook University.

Meyer-Arendt, K.L. (1985) 'The Grand Isle, Louisiana resort cycle', *Annals of Tourism Research*, 12(3): 449–65.

Michener, J. (1951) *Return to Paradise*, New York: Random House.

—— (1962) (orig. edit. 1947) *Tales of the South Pacific*, New York: Macmillan.

Middleton, V. (1989) 'Whither the package tour?', *Tourism Management* 12 (3): 185–92.

—— (1990) *Review of Museums and Cultural Centres in the South Pacific*, Suva: Tourism Council of the South Pacific.

—— (1994) *Marketing in Travel and Tourism* 2nd edn, Oxford: Heinemann.

Middleton, V.T.C. and Hawkins, R. (1993) 'Practical environmental policies in travel and tourism. Part 1: The hotel sector', *Travel and Tourism Analyst* 6: 63–76.

—— (1994) 'Practical environmental policies in travel and tourism. Part 2: The airline and travel sector', *Travel and Tourism Analyst* 1.

Mieczkowski, Z. (1990) *World Trends in Tourism and Recreation*, New York: Peter Lang.

Mignon, C. and Heran, F. (1979) 'La Costa del Sol et son arrière-pays', in A.M. Bernal *et al.* (eds), *Tourisme et Développement Régionale en Andalouse*, 53–133, Paris: Editions de Boccard.

Millar, M. (1992) 'Business as usual while Hamilton looks for buyer', *Travelweek* 27 May: 12.

Milman, A. (1993) 'Maximising the value of focus group research: qualitative analysis of consumers' destination choice', *Journal of Travel Research* 32(2): 61–3.

Mitchell, W.J.T. (1986) *Iconology: Image, Text, Ideology*, Chicago.

Moeran, B. (1983) 'The language of Japanese tourism', *Annals of Tourism Research* 10(1): 93–108.

Moffet, L. (1992) 'Eastwest in Hamilton Island splash', *Australian Financial Review* 2 March: 28.

Moller, H-G. (1983) 'Etude comparée des centres touristiques du Languedoc-Roussillon et de la côte de la Baltique en République Fédérale Allemande', *Norois* 120: 545–51.

Morgan, M. (1991) 'Dressing up to survive: marketing Majorca anew', *Tourism Management* 12(1): 15–20.

Morgan, R. (1993) *Roy Morgan Omnibus Survey: Destination, Awareness and Travel*, Melbourne: Roy Morgan Research.

Morrison, A. (1989) *Hospitality and Travel Marketing*, Albany NY: Delmar.

Mullins, P. (1985) 'Social issues arising from rapid coastal tourist urbanisation', *Australian Urban Studies* 13(2): 19–20.

—— (1990) 'Tourist cities as new cities: Australia's Gold Coast and Sunshine Coast', *Australian Planner* 28: 37–41.

—— (1991) 'Tourism urbanisation', *International Journal of Urban and Regional Research* 15(3): 326–42.

Mumford, L. (1961) *The City in History*, New York: Martin Secker and Warburg.

Murphy, P. (1985) *Tourism: A Community Approach*, London: Methuen.

National Centre for Studies in Travel and Tourism *Queensland Visitor Survey 1992–93*, Brisbane: Government Printer.

National Consumer Council (1988) *Package Holidays: Dreams, Nightmares and Consumer Redress* PD 18/88, London: National Consumer Council.

National Trust for Fiji (1992) *Corporate Plan 1992–1993*, Suva: NTF.

Native Land Trust Board (NLTB) (1980) *A Policy for Tourism Development on Native Land 1980–1985*, Suva: NLTB.

—— (1984) *A Prefeasibility Study of the Development Potential of Malolo Island as a Major Integrated Tourist Resort*, Suva: NLTB.

—— (1990) *A Policy for Tourism Development on Native Land 1990–1995*, 18/90, Suva: NLTB.

Naveh, Z. (1978) 'The role of landscape ecology in development', *Environment Conservation* 5(1): 57–63.

Nayacakalou, R.R. (1975) *Leadership in Fiji*, Melbourne: Oxford University Press.

Niederland, W.G. (1957) 'River symbolism, Part II', *Psychoanalytic Quarterly* 26: 50–75.

Oelrichs, I. (1992) 'Endemic tourism: a profitable industry in a sustainable environment' in *Ecotourism Business in the Pacific, Conference Proceedings*, 14–22, University of Auckland.

Office of the Co-ordinator General (1994) *Draft Whitsundays Tourism Strategy*, Brisbane: Government Printer.

Ogilvie, F.W. (1933) *The Tourist Movement*, London: Staples Press.

Oldmeadow, H. (1992) 'The past disowned: the political and postmodernist assault on the humanities', *Quadrant* March: 60–5.

Olsen, G.W. and Myers, K. (1992) 'Hotels and resorts: family programs and their potential impact', *Hospitality and Tourism Educator* (1): 85–8.

Outridge, P., Torte, D., Ball, J. and Hegerl, E. *The Great Barrier Reef Resource Inventory: Whitsunday Section, Volume 1*, prepared for the GBRMPA.

Pacific Asia Travel Association (1992) *Endemic Tourism: A Profitable Industry in a Sustainable Environment*, Sydney: PATA Think Tank.

Pacific Islands Monthly (1994) 'How fast do government tourist bureaux respond to tourist inquiries?', February: 27.

Pannell, Kerr & Forster (1986) *Conceptual Development Strategy for the Whitsunday Area, Volumes 1 and 2*, prepared for the Queensland Premier's Department.

Parker, S.R. (1983) *Leisure and Work*, London: Allen & Unwin.

Parliament of Queensland (1987) *Integrated Resort Development Act*, Brisbane: Government Printer.

Parry, K. (1983) *Resorts on the Lancashire Coast*, Newton Abbot: David & Charles.

Paul-Levy, F. and Segaud, M.F. (1983) *L'Anthropologie de l'Espace*, Paris: CCI-G Pompidou.

Pearce, D.G. (1978) 'Form and function in French resorts', *Annals of Tourism Research* 5(1): 142–56.

—— (1983) 'The development and impact of large-scale tourism projects: Languedoc-Roussillon and Cancun compared', in C.C. Kissling *et al.* (eds), *Papers, 7th Australian/NZ. Regional Science Assoc.*, 59–71, Canberra.

—— (1987a) 'Mediterranean charters: a comparative geographic perspective', *Tourism Management*, 8(4): 291–305.

—— (1987b) 'Spatial patterns of package tourism in Europe', *Annals of Tourism Research* 14(2): 183–201.

—— (1987c) *Tourism Today: A Geographical Analysis*, Harlow, UK: Longman.

—— (1988) 'Tourism and regional development in the European Community', *Tourism Management* 9(1): 13–22.

—— (1989) *Tourist Development* (2nd edn), London: Longman.

—— (1993) 'Comparative studies in tourism research', in D.G. Pearce and R.W. Butler (eds), *Tourism Research: Critiques and Challenges*, 20–35, London: Routledge.

Pearce, D.G. and Grimmeaud, J-P. (1985) 'The spatial structure of tourist accommodation and hotel demand in Spain', *Geoforum* 15(4): 37–50.

Pearce, D.G. and Kirk, R.M. (1986) 'Carrying capacities for coastal tourism', *Industry and Environment* 9(1): 3–6.

Pearce, P.L. (1977) *The Social and Environmental Perceptions of Overseas Tourists*, unpublished D.Phil. Thesis, University of Oxford.

—— (1981) '"Environmental shock": A study of tourists' reactions to two tropical islands', *Journal of Applied Social Psychology* 11(3): 268–83.

—— (1982) *The Social Psychology of Tourist Behaviour*, Oxford: Pergamon.

—— (1988) *The Ulysses Factor: Evaluating Visitors in Tourist Settings*, New York: Springer-Verlag.

Pearce, P.L. and Moscardo, G. (1987) *Study of the Social Impact of Tourist Boats and Facilities on Existing Users on Norman Reef*, Townsville: GBRMPA.

Perrin, H. (1971) *Les Stations de Sports d'Hiver*, Paris: Berger-Lavrault.

Peterson, K.I. (1987) 'Qualitative research methods for the travel and tourism industry', in J.B. Ritchie and C.R. Goeldner (eds), *Travel, Tourism and Hospitality Research: A Handbook for Managers and Researchers*, 433–38, New York: Wiley.

Pioneer Shire (1988) *Pioneer Shire Development Strategy Report*, North Mackay.

Pizam, A., Milman, A. and King, B.E.M. (1994) 'The perceptions of tourism employees and their families towards tourism: a cross-cultural comparison', *Tourism Management* 15(1): 53–61.

Plange, N.K. (1984) *Tourism: How Fiji People See It and What They Think of It*, Suva: Fiji Tourism Educational Council.

Plant Location International (1993) 'Assessment of social impact of the Club Med proposal at Byron Bay', in *Club Med Proposal. Byron Bay: Assessment of Economic Impacts*, 660–84, Byron Bay: Holiday Villages Pty Ltd.

Plog, S.C. (1974) 'Why destinations rise and fall in popularity', *Cornell Hotel and Restaurant Administration Quarterly* 15: 55–8.

Poon, A. (1989) 'Competitive strategies for a "new tourism"', in C. Cooper (ed.)

Progress in Tourism, Recreation and Hospitality Management 1: 91–102, London: Bell-haven Press.

—— (1993) *Tourism, Technology & Competitive Strategies*, Wallingford: CAB International.

Popcorn, F. (1991) *The Popcorn Report*, New York: Doubleday.

Porter, M.E. (1980) *Competitive Strategy: Techniques for Analysing Industries and Competitors*, New York: The Free Press.

—— (1990) *The Competitive Advantage of Nations*, London: The Macmillan Press.

Portman, J. and Barnett, J. (1976) *The Architect as Developer*, New York: McGraw-Hill.

Potts, T.D. and Uysal, M. (1992) 'Tourism intensity as a function of accommodations', *Journal of Travel Research* 31(2): 40–3.

Pringle, J.D. (1958) *Australian Accent*, London: Chatto & Windus.

Prosser, G. and Lang, J. (1994) 'Postmodern regional economies: tourism and economic development', *National Tourism Research Conference* February, Gold Coast, Qld.

Przeclawski, K. (1992) 'Tourism as the subject of interdisciplinary research', in D.G. Pearce and R. Butler (eds) *Tourism Research: Critiques and Challenges*, 9–20, London: Routledge.

Qalo, R. (1984) *Divided We Stand: Local Government in Fiji*, Suva: Institute of Pacific Studies, USP.

Qantas Airways (1994) *Annual Report 1993–94*, Sydney.

Queensland Government Department of Housing and Local Government (QDHLG) (1991) *Population Projections*, Brisbane: Government Printer.

Queensland Government Tourist Bureau (QGTB) (No date specified) *Bowen and Whitsunday* brochure, Travel Ephemera section, John Oxley Library, Brisbane.

Queensland Tourism Development Board (1947) *Tourist Resources of Queensland and Requirements for their Development*, Brisbane: Government Printer.

Queensland Tourist and Travel Corporation (QTTC) (1993a) *Annual Report 1993*, Brisbane: Government Printer.

—— (1993b) *Corporate Plan*, Brisbane: Government Printer.

Quiroga, I. (1990) 'Characteristics of package tours in Europe', *Annals of Tourism Research* 17(2): 185–207.

Quorum (1986) 'Everything under the sun' 11: 27–9, 31–4.

Rajotte, F. (ed.) (1980) *Tourism as Pacific Islanders See It*, Suva: Institute of Pacific Studies (USP) in association with the South Pacific Social Sciences Association.

Ravuvu, A,D. (1987) *The Fijian Ethos*, Suva: Institute of Pacific Studies, University of the South Pacific.

Redfield, R. (1955) *The Little Community*, Chicago: The University of Chicago Press.

Reilly, R.T. (1982) *Handbook of Professional Tour Management*, Wheaton, Illinois: Merton House.

Reimer, G.D. (1990) 'Packaging dreams: Canadian tour operators at work', *Annals of Tourism Research* 17(4): 501–12.

Rennie, P. (1994) 'Buying a slice of tropical paradise', *Business Review Weekly* 9 May: 72.

Resource Assessment Commission (1992) *Coastal Zone Inquiry. Background Paper*, Canberra: Australian Government Publishing Service.

Richardson, J. (1993) *Ecotourism and Nature-Based Holidays*, Marrickville NSW: Choice Books.

Riesman, D. (1950) *The Lonely Crowd: A Study of the Changing American Character*, Yale: Yale University Press.

Ritchie, J.R.B. (1993) 'New realities, new horizons: leisure, tourism and society in the third millennium', *International Journal of Management and Tourism* 1(3): 311–28.

Ritchie, J.R.B. and Crouch, G.I. (1994) *Competitiveness in International Tourism: A Framework for Understanding and Analysis*, unpublished paper, Calgary: World Tourism Education and Research Centre, University of Calgary.

Ritchie, J.R.B. and Zins, M. (1978) 'Culture as determinant of the attractiveness of a tourism region', *Annals of Tourism Research* 5(2): 252–67.

Ross, G.F. (1990) 'The impacts of tourism on regional Australian communities', *Regional Journal of Social Issues* 25: 15–22.

Roughley, T.C. (1937) *Wonders of the Barrier Reef*, Sydney: Angus & Robertson.

Royal Automobile Club of Victoria (RACV) (1992) *Accommodation Australia*, Melbourne: RACV.

Ryan, C. and Burge, G. (1991) 'Abeles and TNT stumble on the path to grandeur', *Sunday Age: Money* 3, 25 August.

Ryan, J. (1969) *The Hot Land: Focus on New Guinea*, Melbourne: St Martin's Press.

Sack, R.D. (1992) *Place, Modernity and the Consumer's World*, Baltimore: The Johns Hopkins University Press.

Sale, K.M. and Barker, B.C. (1991) *Rock Art on Hook Island: Whitsunday Group North Queensland. Vol. 1: A Recording and Conservation Assessment of the Nara Inlet Rock Art Site*, unpublished report to Queensland Department of Environment and Heritage.

Samy, J. (1980) 'Crumbs from the table? The workers share in tourism', in F. Rajotte and R. Crocombe (eds) *Pacific Tourism as Islanders See It*, Suva: Institute of Pacific Studies and South Pacific Social Sciences Association.

Sandford, J. and Law, R. (1967) *Synthetic Fun*, Harmondsworth: Penguin.

Sandilands, B. (1991a) 'Government hands resorts a lifeline', *Australian Business* 7 October: 19.

—— (1991b) 'How Hayman Island is gnawing into TNT', *Bulletin* 6 August: 102–3.

—— (1992) 'The dream is not over', *Bulletin* 26 May: 27–9.

Sanford, C.L. (1961) *The Quest for Paradise*, Urbana, Ill: University of Illinois Press.

Saunders, J. (1986) 'Ansett to win lease of South Molle Island', *Financial Review* 15 April: 43.

Schneider, P., Schneider, J. and Hansen, E. (1972) 'Modernisation and development: the role of regional élites and noncorporate groups in the European Mediterranean', *Comparative Studies in Society and History* 14(3): 328–50.

School of Travel Industry Management, University of Hawaii (1990) *A Report on Resort Development in Japan*, Manoa: UH Press.

Scott, R.J. (1970) *The Development of Tourism in Fiji Since 1923*, Suva: Fiji Visitors Bureau.

—— (1988) 'Managing in a time of crisis: Fiji after the coups', *Travel and Tourism Analyst*, London: Economist Intelligence Unit.

Seal, K. (1990) 'Child care centers finding their place in lodging industry', *Hotel and Motel Management* 205(21): 1, 41.

See Tho, E.K. (1994) 'Of palm trees and high rises', *PATA Travel News* September: 7–8.

Seers, D., Schaffer, B. and Kiljunen (eds) (1979) *Underdeveloped Europe: Studies in Core–Periphery Relations*, Hassocks: Harvester Press.

Select Hotels and Resorts International (1993) *Select the Natural Pacific*, Sydney.

Serviable, M. *L'Isle Tropicale, Espace Fabuleaux, Espace Fabule*, Voyages Mutualistes, Reunion.

Sewell, S. (1987) 'Moral collapse on Hamilton Island', *Times on Sunday* 11 October: 35.

Sharrad, P. (1990) 'Imagining the Pacific', *Meanjin Quarterly* 49(4): 597–606.

Sheldon, P.J. (1986) 'The tour operator industry: an analysis', *Annals of Tourism Research* 13(3): 349–65.

Sheldon, P.J. and Mak, J. (1987) 'The demand for package tours: a mode choice model', *Journal of Travel Research* 25(3): 13–17.

Siers, J. (No date specified) *The Mamanucas: Islands in the Sun*, Suva: Caines Jannif Ltd.

Silbey, R.G. (1982) *Ski Resort Planning and Development*, Australia: Foundation for the Technical Advancement of Local Government Engineering in Victoria.

Silver, I. (1993) 'Marketing authenticity in Third World countries', *Annals of Tourism Research* 20(1): 302–18.

Sinclair, J. (1987) *Images Incorporated: Advertising as Industry and Ideology*, London: Croom Helm.

Singh, T.V. and Kaur, J. (eds) (1985) *Integrated Mountain Development*, New Delhi: Himalayan Books.

Skinner, J. and Gillam, E. (1992) 'Coastal growth in Northern NSW and Southern Queensland: the facts', in J. Lang (ed.) *Coastal Growth and Management*, Lismore: UNE-Northern Rivers.

Sloth, B. (1988) *Native Legislation and Nature Conservation as a Part of Tourism Development in the Pacific Islands*, Suva: Tourism Council of the South Pacific

Smith, B. (1950) 'European vision and the South Pacific', *Journal of the Warburg and Courtauld Institute* XIII: 66–100.

—— (1989) *European Vision and the South Pacific*, Melbourne: Oxford University Press.

—— (1992) *Imagining the Pacific in the Wake of the Cook Voyages*, Melbourne: Melbourne University Press.

Smith, F. (1994) 'Daydream sale tests interest in tourism', *The Weekend Australian: Property* 10–11 December: 1.

Smith, R.A. (1992) 'Beach resort evolution: implications for planning', *Annals of Tourism Reseach* 19: 304–22.

Sontag, S. (1979) *On Photography*, Harmondsworth: Penguin.

South Molle Island (1993) *South Molle: the All-Inclusive Island Holiday*, brochure valid 1 May 1993 to 30 March 1994.

South, A.K. (1988) 'Classification of accommodation in Australia', in J. Blackwell (ed.) *The Tourism and Hospitality Industry*, 153–60, Chatswood: Australia-International Magazine Services.

Spearritt, P. (1976) *Sydney Since the Twenties*, Sydney: Hale & Iremonger.

—— (1990) 'How we've sold images of Australia', *Australian Society* November: 16–20.

Stallibrass, C. (1980) 'Seaside resorts and the holiday accommodation industry: a case study of Scarborough', *Progress in Planning* 13(3): 103–74.

Stansfield, C.A. and Rickert, J.E. (1970) 'The recreational business district', *Journal of Leisure Research* 2(4): 213–25.

Stansfield, G. (1969) 'Recreational land-use patterns within an American seaside resort', *Tourist Review* 24(4): 128–36.

—— (1978) 'Atlantic City and the resort cycle', *Annals of Tourism Research* 5: 238–51.

Stettner, A.C. (1993) 'Commodity or community? Sustainable development in mountain resorts', *Tourism Recreation Research* 18(1): 3–10.

Stevenson, R.L. (1883) (1948 edn) *Treasure Island*, London: Dent.

Stiles, R.B. and See-Tho,W. (1991) 'Integrated resort development in the Asia Pacific region', *Travel and Tourism Analyst* 3: 22–37.

Stollznow Research Pty Ltd (1990) *Evaluation of Consumer Perceptions of Fiji as a Travel Destination*, Suva: Fiji Visitors Bureau.

—— (1992) *Qualitative Investigation into the Perception of Fiji in Sydney and Melbourne*, prepared for the Fiji Visitors Bureau, Sydney.

Stutts, A.T. and Borsenik, F.D. (1990) *Maintenance Handbook for Hotels, Motels and Resorts*, New York: Van Nostrand Reinhold.

Sunday Age (1994) 'Native title claim on Whitsundays', 4 September: 2.

Sweeney, B. and Associates (1991) *Domestic Tourism Segmentation Study Report on the Islands*, Queensland Tourist and Travel Corporation.

Swinglehurst, E. (1982) *Cooks Tours: the Story of Popular Travel*, Poole: Blandford.

Teare, R. and Olsen, M. (eds) (1992) *International Hospitality Management Corporate Strategy in Practice*, New York: Wiley.

The Economist (1986) 'Club Med: the bourgeois holiday camp', 12 July: 78–9.

Théroux, P. (1992) *The Happy Isles of Oceania*, London: Hamish Hamilton.

Thomas, I. (1993) 'Australian drops Ansett action', *Australian Financial Review* 29 March: 10.

Thompson, G. (1981) *Popular Culture and Everyday Life*, Holidays Unit 11, 5–39, Milton Keynes: OUP.

Thompson, P.T. (1971) *The Use of Mountain Recreation Resources: a Comparison of Recreation and Tourism in the Colorado Rockies and the Swiss Alps*, Boulder: University of Colorado.

Thomson, C.M. and Pearce, D.G. (1980) 'Market segmentation of New Zealand package tours', *Journal of Travel Research*, Fall 1990.

Thurot, J.M. and Thurot, G. (1983) 'The ideology of class and tourism: confronting the discourse of advertising', *Annals of Tourism Research* 10(1): 173–89.

Thurot-Gambier, G. (1981) *Tourisme et Communication Publicitaire*, Thèse de Doctorat de 3ème Cycle, Université de Droit, d'Economie et des Sciences d'Aix-Marseilles.

Times on Sunday (1987) Letter to the Editor, 18 October.

Tindale, N. (1974) *Aboriginal Tribes of Australia*, University of California Press.

Tonge, R. (1987) *How to Conduct Tourism Feasibility Studies*, Coolum, Qld: Gull Publishing.

Tourism Council of the South Pacific (TCSP) (1992) *Vanuatu Visitor Survey 1991*, Suva: TCSP.

Tourism Forecasting Council (TFC) (1995) *Forecast* 1(2), Canberra: Tourism Forecasting Council.

—— (1996) *Forecast* 2(2), Canberra, Tourism Forecasting Council.

Travel Reporter (1992a) 'Solid performance from QTTC', 14 November: 3.

—— (1992b) 'Airlines squabble over Hamilton Island', 14 December: 10.

—— (1993a) 'Comment: too much influence', 18 August: 2.

—— (1993b) 'Government operates in a "policy vacuum" ', 23 July: 4.

—— (1993c) 'It was always done with style and gut feelings', 28 March: 7.

—— (1994a) 'Woodwark Bay construction start up date', 14 January: 21.

—— (1994b) 'Laguna Quays changing trade perceptions', 19 July: 19.

Traveltrade (1994) 'Give wholesalers the whole dollar', 4 May: 25.

Tse, E. and Olsen, M.D. (1990) 'Business strategy and organisational structure: a case of US restaurant firms', *International Journal of Contemporary Hospitality Management* 2(3): 17–23.

Turner, L. and Ash, J. (1975) *The Golden Hordes: International Tourism and the Pleasure Periphery*, London: Constable.

BIBLIOGRAPHY

Turner, V.W. (1984a) *Dramas, Fields and Metaphors: Symbolic Action in Human Society*, Ithaca and London: Cornell University Press.

—— (1984b) *The Ritual Process: Structure and Anti-Structure*, Chicago: Aldine Publishing.

United Nations Development Programme (UNDP) (1973) *A Tourism Development Programme for Fiji*, in association with the Fiji Government and the World Bank, Suva.

Urry, J. (1990) *The Tourist Gaze: Leisure and Travel in Contemporary Societies*, London: Sage.

Usher, L. (1987) *Mainly About Fiji: a Collection of Writings, Broadcasts and Speeches*, Suva: Fiji Times Ltd.

Utrecht, E. (ed.) (1984) *Fiji: Client State of Australasia*, Sydney: University of Sydney Transnational Corporations Research Project.

Uzzell, D. (1984) 'An alternative structuralist approach to the psychology of tourism marketing', *Annals of Tourism Research* 11(1): 79–99.

Valentine, P.S. (1985) *An Investigation of the Visitor Impacts on the Whitsunday Islands National Parks*, unpublished report to Queensland Department of Environment and Heritage.

—— (1992) *Visitor Use in the Southern Whitsunday Islands*, unpublished report to Queensland Department of Environment and Heritage.

Van der Weg, H. (1982) 'Revitalisation of traditional resorts', *Tourism Management* 3(4): 303–7.

Vanclay, F. (1988) *Tourism Perceptions of the Great Barrier Reef*, report to the GBRMPA, Brisbane: Institute for Applied Environmental Research.

Veblen, T. (1899) *The Theory of the Leisure Class*, New York: Macmillan.

Veyret-Verner, G. (1972) 'De la grande station à la petite ville: L'example de Chamonix-Mont Blanc', *Revue de Géographie Alpine* 60(2): 285–305.

Vogt, C., Fesenmaier, D. and MacKay, K. (1993) 'Functional and aesthetic information needs underlying the pleasure travel experience', *Journal of Travel and Tourism Marketing* 2(2/3): 133–46.

Vuidreketi, I. (1994) 'Sun, sea, sand . . . and culture', *Asian Hotelier* March: 10.

Wagner, U. (1977) 'Out of time and place: mass tourism and charter trips', *Ethnos* 42(1–2): 38–52.

Watling, D. and Chape, S. (eds) (1992) *Environment Fiji: The National State of the Environment Report*, Gland, Switzerland: IUCN.

Wernick, A. (1991) *Promotional Culture: Advertising, Ideology and Symbolic Expression*, London: Sage Publications.

Wettlaufer, D. (1973) 'A vision of paradise', in J. Teilhet (ed.) *Dilemmas of Polynesia*, 1–12, San Diego: Fine Arts Gallery.

Wheatcroft, S. (1994) *Aviation and Tourism Policies: Balancing the Benefits*, London: Routledge.

Wheeler, B. (1993) 'Sustaining the ego', *Journal of Sustainable Tourism* 1(2): 121–9.

White, J. (1986) 'Practical implications and requirements for transit, terminal and resort motels', in K. Hollinshead (ed.) *Tourist Resort Development: Markets, Plans and Impacts*, 106–19, Sydney 1986.

Whitsunday Shire Council (1985) *Whitsunday Shire Town Planning Scheme*, Brisbane: Government Printer.

Whitsunday Visitors Bureau (WVB) (1993) *Business Plan 1993 to 1996*, Whitsunday: WVB.

Wilkinson, P.F. (1989) 'Strategies for tourism in island microstates', *Annals of Tourism Research* 16: 153–77.

—— (1994) 'Tourism and small island states: problems of resource analysis, management and development', in A.V. Seaton (ed.) *Tourism: The State of the Art*, 41–51, Chichester: Wiley.

Williams, A.M. and Shaw, G. (1988) *Tourism and Economic Development: Western European Experiences*, London: Bellhaven Press.

Williams, K. (1985) 'Establishing an international resort: a case study', in J. Blackwell (ed.) *The Tourism and Hospitality Industry: 1988–89 Edition*, 189–200, Chatswood: Australian-International Magazine Services.

—— (1988) 'New trends in resort development', *Queensland Planner* 25(2): 14–20.

Witt, S., Brooke, M.Z. and Buckley, P.J. (1991) *The Management of International Tourism*, London: Unwin Hyman.

Witter, B.S. (1985) 'Attitudes about a resort area: a comparison of tourists and local retailers', *Journal of Travel Research* Summer: 14–19.

Wober and Zins (1995) 'Key success factors for tourism resort management', *Journal of Travel and Tourism Marketing* 4(4): 73–84.

World Economic Forum (1992) *The World Competitiveness Report*, Lausanne: IMD International.

World Tourism Organisation (1979) *Presentation and Financing of Tourist Development Projects*, Madrid.

—— (1980) *Physical Planning and Area Development for Tourism in the Six WTO Regions*, Madrid.

—— (1986) *The State Role in Protecting and Promoting Culture as a Factor of Tourism Development and the Proper Use and Exploitation of the Natural Heritage of Sites and Monuments for Tourism*, Madrid: WTO.

—— (1988) *Secondary Tourism Activity Development in Fiji: Opportunities, Policies and Control*, report prepared for the Fiji Government, Suva.

Wright, P. (1985) *On Living in an Old Country*, London: Verso.

Yacoumis, J. (1992) *South Pacific Regional Tourism Marketing Plan*, Suva: TCSP.

Young, J. (1984) *Adventurous Spirits: Australian Migrant Society in Pre-Cession Fiji*, Brisbane: University of Queensland Press.

Young, M. (1992) 'Ecotourism: profitable conservation?', *Proceedings of Ecotourism Business in the Pacific Conference*, University of Auckland, Auckland.

Zann, L. (1992) *The State of the Marine Environment in Fiji. Annexe 3: The Coral Reefs of the Mamanuca Group*, Suva: National Environment Management Project.

INDEX

Printed and bound by CPI Group (UK) Ltd, Croydon, CR0 4YY

01/11/2024

01782629-0016